Science, Politics and Evolution

This book brings together important essays by one of the leading philosophers of science at work today. Elisabeth A. Lloyd examines several of the central topics in philosophy of biology, including the structure of evolutionary theory, units of selection, and evolutionary psychology, as well as the Science Wars, feminism and science, and sexuality and objectivity. Lloyd challenges the current evolutionary accounts of the female orgasm and analyzes them for bias. She also offers an innovative analysis of the concept of objectivity. Lloyd analyzes the structure of evolutionary theory and unlocks the puzzle of the units of selection debates into four distinct aspects, illuminating several mysteries in the biology literature. Central to all essays in this book is the author's abiding concern for evidence and empirical data.

Elisabeth A. Lloyd is Arnold and Maxine Tanis Chair of History and Philosophy of Science at Indiana University and Professor of Biology and an Affiliated Faculty Scholar at the Kinsey Institute for Research in Sex, Gender, and Reproduction. She is the author of *The Structure and Confirmation of Evolutionary Theory*, *Keywords in Evolutionary Biology*, and, most recently, *The Case of the Female Orgasm: Bias in the Science of Evolution*.

Science, Politics and Evolution

ELISABETH A. LLOYD
Indiana University

CAMBRIDGE UNIVERSITY PRESS
Cambridge, New York, Melbourne, Madrid, Cape Town,
Singapore, São Paulo, Delhi, Mexico City

Cambridge University Press
32 Avenue of the Americas, New York NY 10013-2473, USA

Published in the United States of America by Cambridge University Press, New York

www.cambridge.org
Information on this title: www.cambridge.org/9780521684521

First published 2008
First paperback edition 2013

A catalogue record for this publication is available from the British Library

Library of Congress Cataloguing in Publication Data

Lloyd, Elisabeth A.
Science, politics and evolution / Elisabeth A. Lloyd.
p. cm. – (Cambridge studies in philosophy and biology)
Includes bibliographical references (p.) and index.
ISBN 978-0-521-86570-8 (hardback)
1. Evolution (Biology) – Philosophy. 2. Natural selection – Philosophy.
3. Human evolution – Philosophy. I. Title. II. Series.
QH360.5.L55 2008
576.8–dc22 2007027821

ISBN 978-0-521-86570-8 Hardback
ISBN 978-0-521-68452-1 Paperback

Contents

1

The Nature of Darwin's Support for the Theory of Natural Selection

1. DARWIN'S VIEWS

William Whewell and Sir John F. W. Herschel, the most influential writers in philosophy of science in the mid-nineteenth century, both held Newtonian physics aloft as the model form for a scientific theory. In order to demonstrate Darwin's sophistication concerning contemporary philosophical and methodological issues, I shall quote him extensively, in this section, from his private correspondence.

One crucial aspect of the laws of motion insisted on by philosophers and scientists alike was that they could be directly tested or proved. Darwin was well aware of the example provided by Newtonian physics, and was equally well aware that the theory of natural selection could not be tested by direct inference from the evidence.

> When we descend to details, we can prove that no one species has changed [i.e. we cannot prove that a single species has changed]; nor can we prove that the supposed changes are beneficial, which is the groundwork of the theory. Nor can we explain why some species have changed and others have not. (Darwin 1919, 2:210)

On another occasion, Darwin wrote of F. W. Hutton:

> He is one of the very few who see that the change of species cannot be directly proved, and that the doctrine must sink or swim according as it groups and explains phenomena. It is really curious how few judge it in this way, which is clearly the right way. (1919, 2:155)

Lloyd, E. A. (1983). The Nature of Darwin's Support for the Theory of Natural Selection. *Philosophy of Science, 50*, 112–129.

Darwin explained his approach as follows:

> I have always looked at the doctrine of natural selection as an hypothesis, which if it explained several large classes of facts, would deserve to be ranked as a theory deserving acceptance. (1903, 1:139–140)

It is clear from the following letters that Darwin believed his hypothesis to be analogous in this respect to the physical hypothesis concerning light, and accordingly thought that presenting indirect evidence for this theory, as a parallel form of support, should be acceptable scientific practice. He wrote to Asa Gray (1919, 2:80):

> Your distinction between an hypothesis and theory seems to me very ingenious; but I do not think it is ever followed. Every one now speaks of the undulatory *theory* of light; yet the ether is itself hypothetical, and the undulations are inferred only from explaining the phenomenon of light. . . . It seems to me that an hypothesis is *developed* into a theory solely by explaining an ample lot of facts.

Darwin made an explicit comparison in a letter to Henslow (1967, p. 204):

> In a letter to me, [Sedgewick] talks much about my departing from the spirit of inductive philosophy. I wish if you ever talk on [the] subject to him, you would ask him whether it was not allowable (and a great step) to invent the undulatory theory of light – i.e. hypothetical undulations, in a hypothetical substance, the ether. And if this be so, why I may not invent [the] hypothesis of natural selection (which from analogy of domestic productions, and from what we know of the struggle of existence and of the variability of organic beings, is in itself probable) and try whether this hypothesis of natural selection does not explain (as I think it does) a large number of facts in geographical distribution – geological succession – classification, morphology, embryology, etc. – I should really much like to know why such an hypothesis as the undulations of the ether may be invented, and why I may not invent . . . any hypothesis, such as natural selection. . . . I can perfectly understand Sedgewick or any one saying that natural selection does not explain large classes of facts; but that is very different from saying that I depart from right principles of scientific investigation.

Having defended natural selection as a legitimate scientific hypothesis, Darwin supported it with considerations that served either to connect empirical facts to the theory in various ways, or to show that the hypothesis would be useful to biological science in other respects.

Much of Darwin's support for his theory offered in the *Origin* and in correspondence with other scientists consists in linking the theory to empirical evidence in widely varying branches of science. Darwin presented his specific claims regarding his theory in a letter to the journal, *Athenaeum:*

> As far as I can judge, no theory so well explains or connects these several generalizations (more especially the formation of domestic races in comparison with natural species, the principle of classification, embryonic resemblance, etc.) as the theory . . . of natural selection. Nor has any other satisfactory explanation been ever offered of the almost perfect adaptation of all organic beings to each other, and to their physical conditions of life. (1919, 2:207)

Darwin also supported his hypothesis of natural selection by appealing to its value to the science of biology as a whole through promotion of research:

> Whenever naturalists can look at species changing as certain, what a magnificent field will be open – on all the laws of variations, on the genealogy of all living beings, on their lines of migration, etc. (1919, 1:485)

At this point, it may seem quite plausible that Darwin is arguing for his theory chiefly on the basis that it provided the "best explanation" for the different phenomena from widely varying fields. Following a brief summary of the "inference to the best explanation" view of theory choice, I shall argue that such an interpretation of Darwin's theory support is inadequate and misrepresents Darwin's main concerns.

"Inference to the best explanation" has been claimed by Gilbert Harman to be the rule of rational inference basic to all nondeductive inferences (Harman 1965, p. 89). As a rule of inference governing theory choice, inference to the best explanation means that we should infer the hypothesis, which explains a given set of evidence better than any competing hypothesis at hand. Harman's suggestions regarding the criteria used in determining *which* hypothesis is better include simplicity, plausibility, minimal *ad hoc-ness*, and ability to explain a larger quantity of evidence.

Paul Thagard has claimed that Darwin's support for his theory of natural selection is a clear-cut example of an argument to the best explanation; he claims that Darwin supported his theory chiefly on the basis that its explanations were consilient, simple, and analogical in nature, and thus fulfilled some of the prime criteria used in evaluating explanatory theories (Thagard 1978, p. 89). Thagard emphasizes that consilience, simplicity, and analogy (and explanation in general) are primarily concerned with the actual *use* of the theory within a specific social and historical context.

Context-centered features of a theory should be seen in contrast to other features, such as its logical structure (logical relations inside the theory), or its semantic relations (the relations between the theory and the facts). I shall show that although Darwin did use arguments from consilience, simplicity, and analogy to support his theory, his main defense did *not* amount to claims concerning its explanatory power but, rather, to claims regarding its semantic properties. It should *not*, therefore, be considered a clear-cut case of an inference to the best explanation, nor be used as support for that view of theory choice.

2. DARWIN'S SUPPORT

Before proceeding to the actual discussion of Darwin's support for his theory, I would like to introduce an account of the structure of the theory which I will use throughout the rest of the paper. Morton Beckner has argued that natural selection theory is best understood as presenting a family of related models (1959, p. 160). A particular model used in an explanation is constructed by using principles of the theory in combination with certain assumed conditions; the outcome of the model is supposed to be consistent with observed empirical phenomena. A good example is the case of the emergence of two types of wolves, in which Darwin describes an imaginary situation wherein the outcome would be of a certain (testable) kind, given natural selection and granting certain conditions (1964, p. 90).

Particular models serve to show that the theory is compatible with observed phenomena, viz., empirical generalizations and particular facts of evolution, by providing a specific way in which the theory plus certain specific assumed conditions would produce a fit with the empirical results. This account of theory structure, called the "semantic" view, is a recently developed alternative to the so-called received view or covering law approach (see van Fraassen 1972, especially pp. 304–306; van Fraassen 1970, pp. 328–329; Beatty 1980, pp. 399–401, 419–420). Many interpretations of Darwin's support for his theory, including those by Ruse and Kitcher, are guided by the assumption that "good" or "scientific" theory structure must conform to the covering law view (Kitcher 1981, p. 509; Ruse 1979, pp. 109, 236, 270; see especially Ruse 1975a, pp. 221–224). It has been a special problem of the received view to search for empirical laws in theories under analysis; the semantic view circumvents this problem (see Beatty 1980). Throughout the rest of the discussion, I shall rely on a simplified semantic account of natural selection theory as a standard and unifying picture of the structure of the theory. Details concerning the use of natural selection theory in models will be discussed in Section 2.4.

2.1. Consilience

Darwin, as quoted in Section 1, explicitly defended natural selection theory on the basis that "it explained several large classes of fact" (1903, 1:139). I would like first to consider the view that Darwin's claim to consilience is best interpreted as a claim regarding the theory's ability to produce explanations of a certain sort. Following a brief criticism of an explanation-based view of Darwin's claim, I shall suggest an alternative interpretation that grounds consilience instead in the semantic properties of the theory.

According to William Whewell, who coined the term, "consilience" can occur either when an hypothesis can explain at least two known classes of fact, or when it can predict or explain cases of a new and different kind from those cases considered in forming the hypothesis (Laudan 1971, p. 371; Butts 1977, p. 74). Note that these classes are not to be understood as arbitrary sets; evaluation of consilience presupposes a division of facts into natural classes. Whewell's notion has been adopted by Thagard, who claims that consilience can indicate *how much evidence* a theory explains (Thagard 1978, pp. 79–80). This notion can be used in a comparative way; for example, if Theory A explains three classes of facts, and Theory B explains four classes of facts, including the three classes explained by Theory A, then Theory B should be considered more consilient than Theory A. Since this definition is couched in terms of explanations, it seems that consilience is based upon the performance of the theory in explanation. Throughout this discussion, the term "explain" is used in a substantive sense, in which it is contrasted to mere description, prediction, fitting, or accounting for.

Consider, however, the specific support Darwin gave for his claim to consilience: despite his (quite ordinary) use of the word "explain," it does *not*, in fact, involve the level of complexity of an explanation-centered criterion. The use of the word "explain" is nontechnical: it can be replaced by a more neutral term like "account for" without loss to the support given. He claimed that his theory explained "several large classes of fact" (see Darwin 1919, 2:13; 1903, 1:13); these explanations produced with natural selection theory were of a certain type and quality. Given that natural selection can be understood as presenting a family of related models, the form of an explanation using the theory must be a demonstration of how the observable phenomena in question *could* have happened, given moves authorized by the theory plus certain conditions. One evaluates a model by independently testing the assumed conditions and by comparing the outcome of the model with empirical observations, including observations not available at the time of model construction, if possible. In the case of natural selection theory, then, consilience of inductions would consist of the many models one has constructed, using the theory, which have been shown to be at least minimally empirically adequate. Thus, if consilience is defined strictly in terms of explanations, Darwin has not really defended his theory on the basis of consilience. If, by contrast, consilience is seen as based on the relationship or fit between the theory and empirical data, we can make better sense of Darwin's actual defense, as follows. At the heart of Darwin's declarations of the importance of the consilient nature of natural selection theory was his conviction that the theory had proven capable of "saving" the phenomena. In other words, Darwin had shown himself, in the course of his researches, and using his strictest testing criteria, that the theory could be used to construct outlines of models that would give results that *conform to the empirical data.*

The semantic nature of the criterion of consilience is a problem for those who support inference to the best explanation as the rule for theory choice. Those who argue that many scientists have supported their theories on grounds of consilience are correct; the question is whether this support is based on the fact that their theory *explained* the most (in a substantive sense of "explain") – consilience seems to be an index of this sort – or whether consilience is seen instead as an index of the empirical adequacy of the theory, as argued in the preceding paragraphs. If the latter is the case, as I claim, then the inference being made and supported by consilience is *not* an inference to the best explanation, but rather an inference to the most empirically adequate theory.

The fact that Darwin supported his theory by claiming that it was consilient seems to indicate that he was concerned with the "explanatory power" or success of his theory. Because the notion of consilience is basically semantic, Darwin's defense can be more appropriately recognized as a set of claims about the fit of empirical data to various models constructed with the use of natural selection theory. The particular way in which Darwin tested his models, and thereby his theory, is the subject of Section 2.4.

2.2. *Simplicity*

Simplicity, an important constraint on consilience, has been characterized as a function of the number and type of auxiliary hypotheses used in explanations provided by the theory under consideration (Thagard 1978, pp. 85–86). This apparently amounts to evaluation of the use of *ad hoc* hypotheses in explanation – an *ad hoc* hypothesis being one that is only good for explaining the phenomena which it was introduced to explain. I shall first reject the appropriateness of applying a criterion of simplicity of this sort to Darwin's argument. Then I shall discuss Darwin's claim that his theory was valuable because "so many phenomena can be thus grouped together and explained" (1903, 1:184), a claim that seems to involve *some* notion of simplicity.

Consider the nature of explanations from natural selection theory: they consist of models that include the basic theory plus certain empirical hypotheses. For example, an explanation of the present populations of mosquitoes in a certain area would include basic assumptions regarding the fitness of the present occupants' ancestors, in addition to details concerning the availability of food sources for the different species of mosquitoes present. An explanation that includes the additional hypothesis, "there have been no new species of food source plants introduced into the area during the last 200 years," could, when considering the question of the mosquitoes' present food source, make a *better* explanation, that is, one more informative and more persuasive, than an explanation which omits this extra assumption. Thus, in natural selection, explanations are often *improved* by the addition of untested assumptions that were introduced only to explain the phenomenon in question. These empirical assumptions have a strong ring of *ad hoc-ness* until further research reveals more information about them. The initially plausible definition of simplicity based on the number of *ad hoc* hypotheses in the explanations seems therefore completely inappropriate to apply to natural selection theory.

There is, however, a certain simplicity to natural selection theory to which Darwin appeals explicitly. I would like to suggest an alternate formulation of the notion of simplicity, intended to capture the characteristic by which Darwin actually supported his theory.

The set of related models that comprise natural selection theory could more accurately be called a group of model types. Van Fraassen has defined "model type" as the description of a structure in which certain parameters are left unspecified; "model" refers only to specific structures "in which all relevant parameters have specific values" (and van Fraassen has suggested that in its typical use by scientists, "model" has the sense of "model type") (van Fraassen 1980, p. 44). When natural selection theory is said to present a set of related models, it is meant that there are certain model types which are given in the theory to account for observed phenomena; the variables of these model types are specified and instantiated through hypothesis and testing in a recursive manner. For example, a model type used to explain the presence of certain instincts would contain variables corresponding to the possible range or variation of the instinct, its profitability in different sets of circumstances, and the economies of related behaviors, in addition to standard natural selection assumptions concerning the existence of fine gradations of instinct and change in instinct.

A slightly different model type might be used to explain the predominance of one species over related species. Such a model type would include variables representing each species' method of obtaining food, changes in the food supply, flexibility with regard to food supply, and climatic changes, in addition to basic natural selection assumptions regarding the effects of competition and the existence of natural variation. These model types serve as formats for explanations; the particular terms or factors in a model type vary in each application, depending on the outcome of the model and various assumed conditions.

One way to evaluate simplicity of a theory that consists of a set of models is to see how many separate models and terms are necessary to account for the phenomena; roughly, a simpler theory has fewer model types. If a theory is both *simple* in the manner defined above, that is, consists of only a few basic model types, and is *consilient* as well, then it can be considered to *unify* the phenomena. Kitcher has recently given a similar interpretation which, except for its syntactic presentation, presents a similar view of the nature of the unification involved in Darwin's theory: "a theory unifies our beliefs when it provides one (or more generally, a few) pattern(s) of argument which can be used in the derivation of a large number of sentences which we accept" (1981, p. 514). Darwin often supported his theory on the basis of its unifying nature (cf. quote p. 113). This virtue is not one of consilience alone; it also involves the fact that the basic plans for explaining so many classes of phenomena are included in a relatively straightforward and small set of model types.

In other words, the manner in which natural selection theory is simple is that it consists of only a *few* interrelated model types, yet these model types can be filled in with contingent facts and assumptions for *any* specific phenomenon in the extremely large range of phenomena the theory claims to account for. This ability of outlines of evolutionary explanations to be instantiated with different details is also emphasized by Kitcher as a vital aspect of the success of the theory: "[an] eventual unification would consist in derivations of descriptions of these phenomena which would instantiate a common pattern" (1981, pp. 514–515). Thus, simplicity is a relation between the theory, the empirical data, and the explanations constructed from the theory; it cannot, therefore, be considered a relation concerned exclusively with explanation.

2.3. Analogy

As a form of theory support, analogies are generally supposed to improve the explanations provided by the theory. Such support seems to be heuristic support: take the case in which an analogy between phenomena improves the explanations because the first explanation provides a model for the second; the nature of this improvement is an improvement in understanding. Analogy is then playing a heuristic or psychological, rather than confirmatory, role in the justification of the theory.

Darwin's use of artificial selection in support of natural selection is usually brought forth as an example of analogical support; the "familiarity" of artificial selection is supposed to increase the explanatory value of natural selection theory (Thagard 1978, p. 91). Darwin did claim that "the belief in Natural Selection must at present be grounded entirely on general considerations including the analogy of change under domestication by man's selection" (1919, 2:210).

The strategy of Darwin's presentation, wherein the first two chapters of the *Origin* concern artificial selection and abundance of variation in wild organisms, respectively, suggests that he was aware of an heuristic use of the analogy. The situation becomes less clear-cut, however, when we consider that contemporary popular and scientific opinion held that results from domestic variation and artificial selection were strong evidence *against* the transmutation of species (Ruse 1979, pp. 177, 203).

Michael Ruse has argued, in the context of a discussion on the hypothetico-deductive nature of Darwin's theory, that "Darwin's discussions of artificial selection and of variation do have a justificatory role in his thought" (Ruse 1975a, p. 226). Ruse argues as follows: in the course of arguing for natural selection theory, Darwin claims that there *do* exist favorable and injurious variations, on which populational and environmental demands can then exert differential pressure. What evidence does

Darwin give for this assumption, crucial to natural selection theory? He argues from artificial selection: because variations that are useful to people have occurred in domestic animals, it should not be "thought improbable" that variations that are useful to the organism itself in "the battle of life" should occur spontaneously in wild organisms (Darwin 1964, pp. 80, 467).

The assumption that there are heritable variations useful to the (owner) organism should not be confused with the class of individual empirical assumptions that appear in particular applications of natural selection theory. The analogical argument serves as *supporting evidence* for the basic construction of the model types; after this assumption has been made, it becomes possible to assume the existence of *some* heritable favorable variation. The task is then to determine its nature through functional analysis or other research methods. The use of the analogy from artificial selection as supporting evidence for an assumption common to all natural selection model types clearly outstrips a merely heuristic application. Darwin does not use analogy just to improve the explanations by improving understanding (a use which has a basically psychological appeal), nor does the explanation in artificial selection simply serve as a model for explanation in natural selection; rather, analogy seems to be used by Darwin chiefly as the strongest type of evidence by which he could support a certain assumption of his theory (see Ruse 1975a, p. 235; Ruse 1979, pp. 177–178, 203–205).

One last point: it could possibly be misleading to accept Darwin's own interpretation that the supporting role of artificial selection in natural selection theory is analogical. Van Fraassen has suggested that the case of artificial selection might have been serving as a specific, narrow case of a certain process; this process was then generalized from a limited, artificial environment to the natural environment (personal communication).

2.4. Independent Testing of Model Assumptions

The preceding sections contained several references to the fact that Darwin was concerned with testing his theory and its adequacy to empirical facts. I have found that Darwin's concern for consilience was based on his conviction that his models satisfy at least minimal conditions of empirical adequacy, and that theory support criteria that focus on characteristics of explanations neglect the type of argument, significant in terms of both volume and detail, which relates the theory to empirical data and observations.

Once a model is constructed that accommodates the phenomenon to be explained, we can test the theory only by eliminating the various assumptions of the model, that is, by *independently testing* its assumptions. A great deal of the first half of the *Origin* is devoted to providing empirical evidence for the fundamental assumptions implicit in the model types used in natural selection explanations. For instance, Darwin needed some evidence for his assumption that wild organisms spontaneously developed heritable variations that would be advantageous to their survival and reproductive vigor. Darwin needed to be able to take the existence of useful variations for granted; such a general assumption was necessary to create a variable in the model type, which would then be instantiated in the construction of specific models.

The second half of the *Origin* can be understood as a collection of specific models constructed using the model types provided in the first half of the book, in combination with various specific assumptions. Darwin presents these models as explanations for phenomena already publicly known; the support for his theory consists of offering empirical confirmation for all of the special and specific assumptions that are needed, in combination with natural selection, to explain the phenomena. I shall present two cases in which Darwin offers empirical evidence for an empirical assumption of a model constructed with the use of selection theory.

Darwin argues in the *Origin* that one can explain the facts of the geographical distribution of organisms by using natural selection theory. In his famous set of examples involving the Galapagos Archipelago, Darwin hypothesizes that the fact that species inhabiting the islands resemble South American fauna and flora very closely, but are not identical to them, is a result of the migration of ancient South American forms to the islands and their subsequent transmutation through natural selection. In this explanation, the migration of all of the ancestral forms onto the islands is not *directly* testable. It should be proven, however, that such a migration would have been *possible* in order for the model to serve as indirect support for natural selection theory. Accordingly, Darwin discusses "means of dispersal" at length, defending his assumptions regarding migration with a multitude of various observations and experiments (1964, pp. 346–410).

On another occasion, Darwin described in a letter how he created an explanation and then tested an assumption of his model:

> There is a very curious point in the astounding proportion of Coleoptera that are apterous [wingless]; and I think I have guessed the reason, viz.,

> that powers of flight would be injurious to insects inhabiting a confined locality, and expose them to be blown to the sea: to test this, I find that the insects inhabiting the Dezerte Grande, a quite small islet, would be still more exposed to this danger, and here the proportion of apterous insects is even considerably greater than on Madeira proper. (1919, 1:404–405)

The apterous insect instance exemplifies a certain technique for independently testing assumptions: the assumption being considered must be isolated from the other assumptions in the model; this is done by comparing two or more models that have the assumption under question in common; through comparing these models with yet other models, the new assumption can be isolated as the *only* unestablished assumption in the models. This technique (theoretically) makes possible rigorous, though indirect, testing of each assumption in any natural selection model.

The two examples presented here are typical of the empirical support that Darwin cites as indirect evidence for natural selection theory. This evidence serves to confirm the individual assumptions that are necessary to the various specific models constructed when one uses natural selection theory.

2.5. Stimulation of Research

Concern with confirmation of empirical assumptions of the theory is outside the realm of explanation-centered theory evaluation schemes such as inference to the best explanation. Darwin himself emphasized the utility of natural selection theory in areas having little to do with the theory's performance in explanations; furthermore, he saw its success in research-related areas as support for somesort of acceptance of the theory. Darwin supported his theory numerous times, both in private correspondence and in the conclusion of the *Origin*, on the basis of its usefulness in scientific research (1903, 1:104; 1919, 1:485). He argued that "When we regard every production of nature as one which has had a history," a vast field of inquiry would be opened up, involving the causes and laws of variation, the affects of physical conditions, the means of migration, taxonomy based on rudimentary organs and embryology, and more (1964, pp. 485–489). Darwin believed that natural selection theory would be adopted because young scientists would find "that they can group facts and search out new lines of investigation better on the notion of descent, than on that of creation" (1919, 2:147).

Although this makes it sound as if success in serving as a foundation of a research program involves only the relation of the theory to its users, I shall argue briefly that this success is based for a large part on the structural attributes of the theory.

There are four different ways in which natural selection theory can stimulate research. First, the types of explanations generated by the theory serve especially well as outlines for specific research programs. That is, any account of the action of natural selection allows for the possibility of many specific influences, for example, predation, disease, food shortage, severe climate, and so on. An explanation or model of a particular phenomenon must take into account the varying effects of the numerous possible causes of selection pressure. Because of the nature of an explanation using natural selection – that is, that the phenomenon at hand is the *result* of certain assumed conditions plus natural selection – research is always necessary to fill in explicitly specified details during the construction of the model.

Second, natural selection theory provides a strategic plan and order for carrying out different types of research. This research might consist in eliminating alternatives, for example, leading to an hypothesis that selection pressures would most likely be in a particular direction. Once a particular model has been constructed, further research is necessary to confirm or disconfirm the assumptions or hypotheses. Many of these hypotheses, in modern terminology, are basically functional descriptions and analyses.

Third, the technique of isolating and testing the various assumptions of a given model stimulates the acquisition of new empirical information. The ideal situation for a model is when all of the details and assumptions have been rigorously tested independently and confirmed. This can be done by using the isolating technique described in the previous section and exemplified in Darwin's apterous insect example. Use of this technique itself actually stimulates research, because if an assumption is found to be disconfirmed, a new model must be constructed on the basis of different assumptions. One result of this type of research is clearly the acquisition of a great deal of empirical data stemming from the necessity of confirming each assumption.

Finally, natural selection theory stimulates research by allowing the formulation of weak predictions. Researchers can construct a natural selection model based on (mostly) confirmed biological hypotheses, in which the outcome or result of the model has *not* been observed or empirically confirmed. In this case, the task of scientific researchers consists in directed searching for an entity or situation which is within the range of properties predicted, using natural selection theory (see the electric fish example later in this chapter).

Research promoted by natural selection theory can thus take various forms. If the explanations using natural selection are in the form of models that exhibit results compatible with available empirical data, the research program would consist in *constructing* specific models for specific phenomena, independently *testing* the assumptions in each model, and then *comparing* different models. Researchers can also generate models that *predict* the existence, and sometimes general form, of a certain entity or relation; the research suggestions resulting from this type of model would be directions for a real search.

I have outlined several of the ways in which the theory of natural selection is especially well equipped to promote and support a scientific research program. Success of this sort certainly depends on the relation between the users and the theory; but this relation is not the source of its success and acceptance. Rather, the structural characteristics of the theory, for example, its formulation in model types, are a primary source of its success in research.

2.5.1. EXAMPLES OF RESEARCH GUIDED BY DARWIN'S THEORY Darwin himself made use of natural selection theory as a guide for research in his taxonomic work on barnacles. He discovered that, in a certain species of barnacle, there were both male animals and hermaphroditic ones, the hermaphrodites having two tiny "supplemental males" attached on the outside. Darwin comments, "I should never have made this out, had not my species theory convinced me, that an hermaphrodite species must pass into a bisexual species by insensibly small changes; and here we have it . . . " (1903, 1:65).

This is the rough outline of a prediction using natural selection theory. Darwin expected there to exist organisms of a certain approximate description because of certain assumptions in his theory. Although, in this case, he did not actually search for them, he seems to have "kept an eye out" for entities of the appropriate type (within bounds of the prediction). Notice also that Darwin makes the further claim that the phenomenon would not have been identifiable or explainable *unless* he had used a model type of his theory.

Another interesting incident resulted from Darwin's comment in the *Origin* that the electrical organs of fish provided a particularly difficult case for natural selection theory to explain (1964, p. 192). Darwin points out that "if the electric organs had been inherited from one ancient progenitor thus provided, we might have expected that all electric fishes would have been specially related to each other" (1964, pp. 192–193). But electric organs appear in only a few fishes, which are *not* closely related to each other. Thus, natural selection theory seems to suggest a general model which is disconfirmed empirically.

In 1860, a researcher in Dublin, R. McDonnell, wrote to Darwin that the difficulties were even worse; the electric organs were near the head in some fish, near the tail in others, and were supplied by completely different sets of nerves. McDonnell reported (to Darwin) that he realized "that if Darwin is right, there must be homologous organs both near the head and tail in other non-electric fish" (Darwin 1919, 2:145). McDonnell's subsequent search for these organs was successful, and he published his findings.

There are two ways to interpret this incident; one emphasizes model construction, whereas the other highlights its predictive aspect. When Darwin considered the problem, he seems to have constructed a model something like this: there was an electric fish that was the common ancestor of all the present-day electric fish (assumption); by inheritance and speciation, we would expect the existence of several species of electric fish with close affinities (outcome of the model type). As Darwin pointed out, under existing evidence, the initial assumption was disconfirmed empirically. McDonnell's approach was to turn Darwin's model around: there is a common ancestor of all electric fish (assumption); by inheritance and speciation, it produced the several species of electric fish (assumption); there *must be* some minimal morphological resemblance among all descendents of the ancestor fish; in particular, the electric organs must have homologues in the nonelectric fish, which are closely related to the electric ones *(prediction)*. Note that the type of predictions that natural selection theory can produce are comparatively weak; in any predictive case, only a range of characteristics or possibilities can be specified – precise predictions involving *which* element in this range will be present are not supportable. That McDonnell's general empirical prediction was vindicated appeared to Darwin as strong support for this theory (1919, 2:145).

I take my final examples from the field of palaeontology. Darwin's theory exerted enormous influence on palaeontological research, some of which led to spectacular empirical discoveries, especially in the United

States (Pfeifer 1972). I can merely hint at the importance of these discoveries and their use as evidence for natural selection theory. Two related assumptions seem to serve as the basis for the most dramatic and well-known discoveries made with the use of natural selection theory: (1) given any two species, there existed some common ancestor at some point in the history of the world (1919, 2:121); and (2) given two species that occurred in different epochs but are related on the generic or familial level, it can be assumed that any species intermediate *between* the two species would also occur at a *time* intermediate between the two species (Darwin 1964, p. 462). These assumptions are confirmable through palaeontological evidence.

Albert Gaudry, a French palaeontologist, constructed tree diagrams presenting his evidence for the evolution of various mammalian families, and by doing so, established that new discoveries tended to *fill in* the gaps between previously known species and genera (Rudwick 1976, p. 240). Gaudry also showed that the forms discovered which were intermediate *anatomically* between the horse genus *(equus)* and other odd-toed ungulates were also intermediate in the geological time scale, thus supporting the second assumption. Confirming evidence of the first assumption mentioned above was provided by the discovery in 1861 in Bavaria of *Archaeopteryx*, a fossil bird with unmistakable reptilian traits. From the same location, *Compsognathus* was discovered, a reptile with birdlike anatomical features. Both of these fossils were argued by Thomas Huxley to provide "missing links" through showing that distinct classes of animals have had common ancestors (Rudwick 1976, pp. 250–251).

Subsequent palaeontological research continued to serve evolutionary theory in two ways. If one assumes that natural selection theory is a good theory, then there is a certain result that must not be disconfirmed, that is, that lineages revealed by fossil remains conform to the expected temporal and anatomical succession. That palaeontological discoveries consistently produced this result served as strong confirmation of a fundamental assumption of all natural selection model types. These examples demonstrate several related results from research promoted by natural selection theory. The testing of assumptions in various specific models led to new empirical findings in various fields, including classification, comparative anatomy, and palaeontology. Furthermore, by assuming specific assumptions when constructing models, researchers were able to predict the existence and general features of previously undiscovered entities or relations. The testing of assumptions and predictions served to focus and guide research efficiently into specific programs.

3. EVALUATION OF SUPPORT

Darwin thought that he had no direct evidence for natural selection equivalent to the supposedly direct evidence offered by Newtonians for the laws of motion, the paradigm of scientific theory at the time. He moved the fight into a more appropriate arena by drawing a parallel between the hypothesis of natural selection and the hypothetical wave theory of light. As a result, any demands for direct evidence for natural selection theory made by Darwin's contemporaries were countered by the claim that perfectly good and accepted scientific theories could be supported by indirect evidence and support of various kinds. Darwin accordingly provided a vast amount of indirect evidence in the form of support for the various assumptions necessary for model types and particular models constructed from natural selection theory. Considered in this context, Darwin's evidence cannot he considered inferior either by the standards of his contemporaries or by today's.

A major objection, raised against Darwin ever since 1860, is that natural selection theory does not make testable predictions. I have discussed several examples of predictions resulting from the application of natural selection theory to model construction. Mary Williams has recently argued that modern evolutionary theory yields testable predictions, supporting the status of natural selection theory, a subset of modern evolutionary biology, as a predictive theory (1982, p. 304). In addition, natural selection models constructed from confirmed assumptions led to specific successful research programs. The success of the research lay not just in the fact that new empirical information was acquired; the new facts also indirectly *confirmed* the models and the theory. In the case that the new facts were of the sort suggested by a model constructed using natural selection theory, the success of the research must therefore be closely related to the predictive success of the theory.

Darwin's specific support of his theory of natural selection consists of various sorts of arguments. Several of these, particularly his claims to consilience, simplicity, and analogy, seem, *prima facie*, to be aspects of the explanatory use of the theory. On closer examination, though, Darwin's claim that his theory was highly consilient actually amounts to the claim that the theory had exhibited empirical adequacy in a large variety of instances, according to all available data. The notion of consilience is therefore not primarily a characteristic of explanations; rather, it is a semantic relation, based on the relation between the theory and the empirical data.

In addition, a criterion of simplicity that focuses on the form or content of the explanation is not an element of Darwin's support for natural selection. The simplifying and unifying aspects of the theory to which Darwin *does* refer are not merely relations between the theory and users but involve essential relations of the theory to empirical facts. Considering the analogical inferences in Darwin's theory as merely heuristic devices does not accurately represent their role in Darwin's argument. The existence of heritable variations is an assumption basic to all model types representing natural selection, and this assumption is supported by analogical evidence, and possibly generalization from a limited to a total environment. Analogical evidence is a semantic issue, and does not directly concern various features of explanation, as assumed by proponents of an inference to the best explanation view of Darwin's defense.

Aside from the fact that explanation-centered criteria are inappropriate for characterizing the nature of the above aspects of Darwin's support for his theory, such criteria omit a crucial line of his defense: the role of the theory in stimulating scientific research. Thus, the structure of natural selection theory, regarded as presenting a set of related model types, makes it particularly suited to serve as a foundation for scientific research. Research generated by the theory, according to my simplified version, serves either to test assumptions of particular models or to search for or identify entities or relations which are predicted by speculative models.

I have summarized some instances of the vindication of predictions that were supplied by models using natural selection theory. Such successful research also contributes to the support of natural selection theory itself. I conclude that the support offered by Darwin and other nineteenth-century researchers is *not* inherently weaker than, or of a kind inferior to, support offered for other scientific theories. Darwin supported the assumptions contained in the model types of natural selection by extensive empirical evidence. Under the covering law approach taken by Ruse, Kitcher, and others, interest is focused on Darwin's formulation of various "empirical laws" that can be fitted together into a partial axiomatization of the theory of natural selection (see especially Ruse 1975a). In contrast, the conception of natural selection theory as presenting a family of related model types suggests an interpretation of Darwin's argument that emphasizes and clarifies the role of his empirical evidence.

Explanatory power either involves relations strictly between the user and the theory, or among the user, theory, and data; theory structure and empirical adequacy, by contrast, are relations founded in the logical structure of the theory itself and the fit between empirical data and theory – relations that are much less context-oriented than those heavily involving users. Contrary to the claim that Darwin supported his theory chiefly on the basis of its explanatory power, I have argued that the structure of the theory and its relation to empirical evidence, as shown both in explanation and in research, served as Darwin's primary support for his theory.

2

A Semantic Approach to the Structure of Population Genetics

1. INTRODUCTION

Known to be unlike Newtonian mechanics but also unlike Creationist biology, modern evolutionary theory has a structure that has proved difficult to characterize. Recently, John Beatty (1980, 1981, 1982) and Paul Thompson (1983) have approached the problem of describing the structure of evolutionary theory using the semantic approach developed by P. Suppes, B. C. van Fraassen, and F. Suppe.

Advantages of the semantic view over the logical positivist approaches that, until recently, dominated discussions of theory structure, have been presented by Suppes (1957, 1961, 1962, 1967), van Fraassen (1970, 1972, 1974, 1980), Suppe (1972, 1973, 1974, 1977), and Stegmuller (1976), and will not be discussed here.

The semantic approach, in which I include both the "set theoretic" and "state space" approaches, has been used to describe the structure of Newtonian mechanics, equilibrium thermodynamics, quantum mechanics, and parts of biological theory (Sneed 1971; Stegmuller 1976; Suppes 1957; Wessels 1976; Moulines 1975; van Fraassen 1970, 1972, 1974; Suppe 1974a; see Suppe 1974, 1979 for a summary of the semantic view and its literature).

I shall assume the following positions in this paper, without defense: the semantic view – in particular, the state space version – is more suited to a description of evolutionary theory than any axiomatic view (Beatty 1980, 1981, 1982; Thompson 1983); the semantic view can provide a richer and more useful description of the structure of a theory than axiomatic

Lloyd, E. A. (1984). A Semantic Approach to the Structure of Population Genetics. *Philosophy of Science*, *51*, 242–244.

approaches (see especially Suppe 1972; van Fraassen 1980); and, finally, the semantic view is capable of formally describing theories that are not describable using any axiomatic approach (van Fraassen 1986).

My goal in this paper is to provide an introduction and further development of the semantic approach to the structure of population genetics theory, the most formal and developed subtheory of contemporary evolutionary theory.[1] In the process of providing a formal framework for the detailed description of the theory, I provide a means by which precise analysis of theoretical problems can be carried out. Ultimately, the utility of describing population genetics (and evolutionary theory as a whole) through the semantic view rests on the ability of the semantic view to provide an analytical framework sensitive to the relevant theoretical problems. A working model, that is, an actual semantic description of the theory, must therefore be available before we can evaluate its power as an analytical tool.

After a brief look at the semantic view in the rest of Section 1, I present a range of population genetics models in Section 2, in order to illustrate the variety and general character of the theory, and the suitability of the semantic approach. In Section 3, I discuss particular problems encountered in describing population genetics models using the semantic view of theory structure as a framework. Promising lines for further research that makes use of this approach are noted throughout the paper.

1.1. The Semantic View of Theory Structure

According to the semantic view of theory structure, a scientific theory specifies certain kinds of systems. The systems specified by a scientific theory are ideal; they are used in scientific explanation through claims that certain systems in the natural world are of the kind defined by the theory. The semantic view offers a formal approach to analyzing these systems, which are usually understood as mathematical structures. There are different ways to describe these structures formally: the set-theoretic predicate

[1] Population genetics theory, because of its advanced formal development, lends itself to analysis by the semantic view. My use of population genetics as a starting place for the analysis of the structure of evolutionary theory as a whole does *not* imply, however, that population genetics is assumed to constitute the "core" or "foundation" of the theory. On the contrary, I assume that population genetics, as a set of structures, is embedded in the larger structure called evolutionary theory. It seems likely that if the semantic approach can be used to describe the most structured segment of the theory, it may provide a good approach to the theory as a whole.

approach, developed by Suppes, Sneed, and Stegmuller, involves description by a set-theoretic predicate; the state space approach, employed by van Fraassen and Suppe, describes the structures as configurations of certain mathematical spaces. In this paper, we will be using van Fraassen's version of the state space approach (see Suppe 1979, for discussion of differences among the various semantic approaches).

In general, a structure presented by a theory (understood as intended to represent empirical phenomena) is a *model* of the theory if it satisfies the theorems of the theory. In a semantic definition, the set of sentences that are theorems of the theory is defined *not* in relation to a set of axioms, but by directly defining the class of structures; for any given language *L*, the theorems of the theory in *L* are the sentences of *L* which are satisfied in all these structures. Reference to syntax or to a syntactically defined set of theorems is thus unnecessary. The models picked out are mathematical models of the evolution of states of a given system, in both isolation and interaction, through time. This selection is achieved by conceiving of the ideal system as capable of a certain set of *states* – these states are represented by elements of a certain mathematical space, the *state space* (van Fraassen 1970, p. 238; 1972, pp. 303, 305). (NB In this paper, "models" and "systems" *always* refer to *ideal* systems; when the actual biological systems are being discussed, they will be called "empirical" systems.) The variables used in each mathematical model represent various measurable or potentially quantifiable physical magnitudes. Classically, any particular configuration of values for these variables is a *state* of the system, the state space or "phase space" being the collection of all possible configurations of the variables.

The theory itself represents the behavior of the system in terms of its states; the rules or laws of the theory (i.e., laws of coexistence, succession, or interaction) can delineate various configurations and trajectories on the state space. Under the semantic view, these structures, "being phase spaces of configurations imposed on them in accordance with the laws of the theory," are themselves seen as constitutive of the theory (Suppe 1977, pp. 226–227).

Description of the structure of the theory itself therefore involves only the description of the set of models presented by the theory. It is crucial, then, to discuss the various necessary components of describing a model.

Construction of a model within the theory involves assignment of a location in the state space of the theory to a system of the kind defined by the theory. Potentially, there are many kinds of systems that a given theory can be used to describe; limitations come from the dynamical sufficiency (i.e., whether it can be used to describe the system accurately and completely) and the effectiveness of the laws used to describe the system and its changes. Thus, there are two main aspects to defining a model. First, the state space must be defined – this involves choosing the variables and parameters with which the system will be described; second, coexistence laws, which describe the structure of the system, and laws of succession, which describe changes in its structure, must be defined.

Defining the state space involves defining the set of all the states the system could possibly exhibit. Certain mathematical entities – in the case of the models we shall be looking at, these are *vectors* – are chosen to represent these states. The collection of all the possible values for each variable assigned a place in the vector is the state space of the system. The system and its states can have a geometrical interpretation: the variables used in the state description (i.e., state variables) can be conceived as the axes of a Cartesian space. The state of the system at any time may be represented as a point in that space, located by projection onto the various axes.

The family of measurable physical magnitudes, in terms of which a given system is defined, also includes a set of parameters. The biologist R. C. Lewontin defines parameters as values that are not themselves a function of the state of the system[2] (1974, pp. 7–8). Thus, a parameter can be understood as a fixed value of a variable in the state space – topologically, setting a parameter seems to amount to limiting the number of possible structures in the state space by reducing the dimensionality of the model (see Section 3.4).

Laws, used to describe the behavior of the system in question, must also be defined in a description of a model or set of models. Laws have various forms: in general, coexistence laws describe the possible states of the system in terms of the state space, whereas changes in the state of the system are described by laws of succession. Suppe has given a complete, formal classification of succession and coexistence laws according to the semantic view (Suppe 1976); detailed discussion of the various evolutionary laws in terms of his system cannot be done here. Rather, we will discuss, in Section 3.3, certain problems of classification encountered in analyzing evolutionary laws.

[2] Lewontin notes that although parameters can involve time and can change *over* time, they are not correlated to the variable value as it changes over time (personal communication).

At this point, I would like to draw a distinction. Consider the problem of determining the most appropriate state space with which to represent genetic changes in populations; this is, to an extent, an *empirical* question. Determination of the types or categories of state spaces *used* in population genetics, however, and the relation of these state space types to determination of the structures comprising the theory, are *philosophical*, rather than empirical, questions.

With this distinction in mind, I examine a few examples of population genetics models in Section 2. My purpose is threefold. First, having presented the general terms in which I propose to describe population genetics models, I illustrate these terms through a few actual models. This is an easy and natural task, since much of the theory is *presented* in these same terms. The second goal, then, is to demonstrate that the state space version of the semantic approach provides a natural reconstruction of the theory – less arbitrary than, for example, an axiomatic approach – because it makes sense of the theory as presented. Third, I hope to show, particularly through the example in Section 2.3, that the semantic approach highlights some features of population genetics theory which are theoretically important. Detailed discussion of the description of population genetics models is presented in Section 3.

2. MODELS IN POPULATION GENETICS

Population genetics, as characterized for example by Lewontin, is the "study of the origin and dynamics of genetic variation within populations" (1974, p. 12). The notion of "gene frequency" is fundamental; description of both changes and equilibria of gene frequencies in populations is a primary goal of population genetics theory.

The Hardy-Weinberg law, an equilibrium law of gene frequencies, serves as the foundation of population genetics theory. Take a single locus (gene) with only two alleles, A and a (alternate types of that gene), in a population of diploid organisms (organisms with paired chromosomes). Take the frequency of allele A to be p, the frequency of a to be q (with $p+q=1$). The Hardy-Weinberg law gives the genotype frequencies of the zygotes (the potential next generation) by the equation:

$$p^2 AA + 2pqAa + q^2 aa = 1$$

The system represented by this equation is a "one-locus" system, that is, calculations are performed assuming the complete isolation and independence of the alleles at each locus. Furthermore, a completely random mating pattern is assumed, that is, the genotype makes no difference to the probability of mating to any given genotype. It is also assumed that each genotype contributes equally to the pool of gametes from which the zygotes are randomly "chosen," but this is not generally the case. The comparative contribution of each genotype to the next generation is its fitness value. More complicated models, involving the individual (w) or population fitness value (w bar), in conjunction with the basic Hardy-Weinberg law, are necessary in order to describe all but the most simplistic system. In the rest of this section, I shall present a few examples of these more complex models.

2.1. Deterministic Models

Consider the case in which carriers of a certain genotype contribute a larger proportion of gametes (reproductive cells carrying only half of the complement of the organism's chromosomes) to the gene pool than the other genotypes (of the locus under consideration). Some modification in the Hardy-Weinberg equation is necessary, because it represents equal contributions from genotypes to the gene pool. The difference in contribution is a measure of the "fitness" or "selective value" (w) of a given genotype. The fitness of the genotype contributing the most is taken by convention as 1; the other genotypes have fitnesses of $(1-s)$, where the value of s is the selection coefficient of that genotype.

In a case of simple dominance, in which the fitness of genotypes AA and Aa equals 1, and the fitness of aa is $(1-s)$, we can predict the frequencies (in the ideal system) of the genotypes after selection through a modification of the Hardy-Weinberg equation:

$$p^2 AA + 2pqAa + (1-s)q^2 aa = 1 - sq^2$$

We can then calculate the frequency p' of the A allele in the next generation:

$$p' = (p^2 + pq)/(1 - sq^2) = p/(1 - sq^2)$$

So the increment, Δp of the frequency of allele A in one generation is:

$$\Delta p = spq^2/(1 - sq^2)$$

(from Dobzhansky 1970, p. 102). If s is very small (0.01 or less), it is possible to calculate analytically an equilibrium value p_E (such that $p_E' = p_E$) for the frequency of A. Calculations of the number of generations taken for a given change in gene frequency are then also possible (Maynard Smith 1968, pp. 74–75; Lewontin 1967, p. 81). This sort of model is a *deterministic* model since, given the initial conditions of the population – in this case the initial gene frequency – and any set of parameters – in this case the selection coefficient – the precise condition of the population at some future time can be predicted (Lewontin 1967, p. 81; see Section 3.3).

More than one parameter can be incorporated into the basic model based on the Hardy-Weinberg equations. For example, mutation rates can be included, so that the frequency in the next generation depends both on selection and on mutation. Take the mutation rate from a to A as μ, where μ is defined as the probability that a has mutated to A within the time of one generation. The frequency in the next generation is calculated as follows (where s is the negative fitness coefficient of A):

$$p' = \frac{2p(1-s) + 2\mu q(1-ps)}{2 - 2s(p^2 + 2pq)} = \frac{p - ps + \mu q - \mu pqs}{1 - sp(p + 2q)}$$

In other words, the frequency of allele A in the next generation is calculated in terms of both parameters, μ and s. Once again, this is a deterministic model, because a definite gene frequency results. The model can be simplified greatly by assuming that p is very small, which is plausible under the assumption that the A allele is deleterious, and hence would be maintained only at low frequency. If p is small, then we can approximate using

$$p' = p - ps + \mu q$$

and at the equilibrium state

$$p_E = p_E - sp_E + \mu(1 - p_E)$$

or, if μ is small relative to s, we can approximate by

$$p_E = \frac{\mu}{s} \quad \text{(Maynard Smith 1968, p. 79).}$$

In general, in deterministic models the initial conditions of the population are represented by an ordered *set* of values of variables, that is, a vector. The above examples used a set of only one variable, p. A parameter set is also specified, μ and s in the previous example; the value for the variable after a certain time interval is given by equations incorporating the parameters. Such equations embody the dynamical *laws* of change for the system; they entail a theory about the equilibrium states of the system.

2.2. Stochastic Models

With some evolutionary processes, a number of different results are possible. The mathematical models must, in these cases, represent the relative chances of the occurrence of each of the possible results. In one example of such a probabilistic or "stochastic" model, the probability that an allele with selective coefficient s will reach fixation (i.e., have frequency of 1) within a population of effective size N over many generations is evaluated. The result of this type of model will be a probability distribution rather than the single value specified by a deterministic model. That is, the model will specify the probabilities of the various possible final states, but will *not* say which one will occur, even if we know *only* one will occur. The model can be understood as having "ergodic properties," that is, at equilibrium there is some final probability $p(1)$ of the system being in state 1, another probability $p(2)$ of the system being in state 2, and so on (Lewontin 1967, p. 81).

Thus, p_s in

$$p_s = \frac{1 - e^{-4Nsp}}{1 - e^{-4Ns}} \quad \text{(Kimura and Ohta 1971, pp. 9–10)}$$

(where p is the frequency of the allele at the beginning of the process) can be understood roughly as the proportion of total populations of effective size N, which, confronted with an allele with selection coefficient s, would eventually reach a frequency of 1 for that allele (i.e., eliminate all other alleles at that locus).

The need for stochastic models arises when it is necessary to know more than the average of a range of values, that is, when we need to measure variability. The basic way to handle essential (i.e., necessary) variability is to use an appropriate probability distribution that represents the chance that an individual selected at random will be found to have any given value or range of values. A number of different types of distributions are useful.[3]

[3] For continuous measurements, a normal (Gaussian) probability curve

$$y = \frac{1}{\sigma\sqrt{2\pi}} \exp - \frac{(x-\mu)^2}{2\sigma^2} (-\infty < x < \infty)$$

provides an accurate representation. The equation is put in terms of two parameters, the mean, and σ: the standard deviation. In application, the likelihood of observing an individual with the character in the range from x_1 to $x_2 = \int_{x1}^{x2} y dx$.

In most stochastic models in population genetics, the biologist attempts to predict the way in which the "ensemble of populations" changes in time, and what the equilibrium distributions look like. This is basically statistical mechanics, and the problems can be solved by borrowing methods from that branch of physics (Lewontin 1967, p. 82). For example, in order to solve the distribution function of the gene frequency at equilibrium, change in the ensemble is often approximated as a partial differential equation in time, though this is not always possible or practical (see Bailey 1967, p. 42).

As the mathematical models used to represent genetical phenomena incorporate more parameters and information – in order to make them match the empirical results more closely – it becomes more difficult to arrive at precise mathematical solutions. Yet there is still a need to formulate the complex models in well-defined mathematical terms (Bailey 1967, p. 43). In cases in which approximations cannot be done, simulations are often used. These simulations are "realizations of a stochastic model which are strictly analogous to possible realizations of a real-life process" (Bailey 1967, p. 43). In other words, a computer is used to produce a large number of artificial (as opposed to actual, laboratory) realizations of the stochastic process in question. A large number of runs are executed, using alternative combinations of the values of the parameters, which are fixed for each particular model. With the collection of model results in hand, the biologist can then compute the means, variances, and so on for the models. Simulation models also can aid future research by providing (1) information about what measurements might be useful, and (2) a means of estimating parameter values (Starfield et al. 1980, pp. 338, 353; Bailey 1967, p. 42).

2.3. *Example*

Lewontin and Dunn's work on polymorphism in the house mouse, discussed later, provides an interesting example of both the simulation of a stochastic process, and an explicit comparison between deterministic and stochastic models.

For discrete variables, a binomial distribution can be used. Taking n individuals for which the (independent) chance of having a certain character is p, the chance of observing r individuals with that character is

$$\frac{n!}{r!(n-r)!}p^r(1-p)^{n-r}(0 \leq r \leq n)$$

(equations from Bailey 1967, p. 25).

Lewontin and Dunn (1960) examined a situation in which the existence of a mutant *t* allele at a specific locus is widespread among the populations studied. The polymorphism (presence of more than one type of allele of the gene) is unusual. Strong selection against its maintenance in the population is assumed because it is lethal when homozygous (except in three cases, in which it causes male sterility). These *t* alleles, however, are also subject to a strong abnormality in the process of gamete production and maturation. Under normal conditions, 50 percent of the gametes of a heterozygote will contain one allele, and 50 percent the other. The heterozygote containing the *t* allele, however, yields an abnormal ratio of 95:5 of *t* to normal gametes – rather than the expected 50:50 – now known to result from differential mortality of the gametes as they mature (see Bennett 1975).

The problem for the biologist is to explain how the polymorphism is maintained in the population.

In general, the presence of the polymorphism is accounted for by a balance between the two forces cited earlier: the selection against the mutant *t* allele in the homozygote reduces the number of such alleles, whereas the stock of *t* alleles is constantly increased by the abnormal gamete ratios in the heterozygous males. Heterosis, that is, superior fitness of the heterozygote, also might serve as a balancing force, but this force is omitted from these models because of lack of data (Lewontin and Dunn 1960, p. 707).

Both deterministic and stochastic models can be used to represent the key features of this qualitative account. The choice, in this case, turns on assumptions about the value of the parameter for population size.

If the breeding groups (i.e., effective population size) are assumed to be small, then chance processes, such as random drift, add a statistical element to the situation, necessitating the use of a stochastic model. That is, Wright showed that if you have a finite population size, the rates of changes in gene frequencies will depend, among other things, on random processes involving mutation rates, migration rates, selection, and "accidents of sampling." One particular result is that it is easier to reach fixation (of an allele) in small populations, in which genes are lost or fixed at random, with little reference to selection pressure (Dobzhansky 1970, pp. 230, 232–234). With small populations, then, the presence of the polymorphism is *not* understood to be purely a function of the interaction of the two forces discussed earlier, so a deterministic model is not appropriate (Lewontin and Dunn 1960, p. 707).

If, by contrast, the population size is assumed to be effectively infinite, then the random effects resulting from small population size are absent, and the state of the polymorphism in the population is solely a function of the selection and abnormal segregation values (although infinite population size is not necessarily required by deterministic modeling).

The deterministic model, chosen first to account for the frequencies of this polymorphism (from Bruck 1957), uses two parameters: the proportion of mutant *t* gametes in the effective sperm pool, m; and *p*, the frequency of the non-mutant allele. The result of this model is a *single value* for the frequency of adults heterozygous for the *t* allele. This result was found not to correspond with the result in nature (Lewontin and Dunn 1960, p. 708).

In addition to the empirical inadequacy of the deterministic model, the biologists had theoretical reasons to believe that a stochastic model would be more appropriate for this phenomenon. That is, they noted that the effective size of a breeding unit is small; the species population as a whole consists of a number of partly separated, relatively small, breeding groups. Thus, Lewontin and Dunn decided that "a useful approach in the construction of models is to test the effects on gene frequencies of small effective size of the breeding unit" (1960, p. 708).

Lewontin and Dunn analyzed the stochastic model of the processes of the interaction of selection, segregation abnormality, and restricted population size by *simulation*. The simulation is done by making rules for the evolution of simulated populations that "conform with genetic rules of meiosis, fertilization, and selection" (1960, p. 708). Random elements are also included in the models, because chance is involved in the survival and reproduction of any particular individual (selection), and in which gametes are chosen from the gamete pool. Randomization of the union of sperm and egg yields different frequencies on each run of the simulation (done on computer, in most cases). The idea is to collect a number of these different frequency results and get a *distribution* of the results over a number of runs (Lewontin 1962, p. 67). The parameters fixed for each run include *N* (effective population size), the fitnesses of the various genotypes, and *m* (the factor of segregation distortion). Each run is started with an exact description of the initial population (Lewontin and Dunn 1960, pp. 708–710).

Large numbers of runs are made with identical parameter sets; no two of these runs will have the same results, because of random factors. Distributions, means, and variances can be calculated from the gene frequency results obtained from all the models with a given parameter set. Lewontin and Dunn's statistical analysis of their simulated results led them to conclude that the effects of changing the population size are statistically significant, that is, use of a smaller value for the population size parameter results in genetic drift. The actual distributions obtained by Lewontin and Dunn from the simulation of the stochastic model conform with the predictions made by Wright's mathematical model (1960, p. 712). They conclude that for small populations, the mean values from the stochastic model do not correspond with the prediction from the deterministic model, because the latter model does not account for the chance loss of alleles in small breeding groups (1960, p. 719).

Thus, with the application of a stochastic model to small breeding groups, it is possible to produce simulation results that fit the actual results better than the deterministic model. Information is also gained regarding the exact inadequacies of the deterministic model for the particular phenomenon being modeled. In this case, the assumption of infinitely large effective population size, N, led to inadequacy of the model containing that assumption.

3. THE STRUCTURE OF POPULATION GENETICS THEORY

Having presented a few particular examples of population genetics models which highlight the presence and utility of certain facets of model description, I would like to discuss details of the description of the theory according to the semantic view. Formalization of any theory T, according to the semantic view, involves defining the class of models of T. The theory is conceived as defining a kind of ideal system. The main items needed for this description are the definition of a state space, state variables, parameters, and a set of laws of succession and coexistence for the system (see Section 1.1). Section 3.1 discusses the most common state space for the representation of genetical phenomena of populations, and its theoretical disadvantages. Choice and evaluation of parameters, and the relations between parameters and the structures, are discussed in Section 3.2. Section 3.3 contains some general comments regarding the laws or rules of the models. Finally, I discuss very briefly the interrelationship among the models.

3.1. State Spaces

Choosing a state space (and thereby, a set of state variables) for the representation of genetic states and changes in a population is an important part of population genetics theory. As Lewontin notes

> The problem of constructing an evolutionary theory is the problem of constructing a state space that will be dynamically sufficient, and a set of laws of transformation [i.e., laws of succession] in that state space that will transform all the state variables. (1974, p. 8)

Paul Thompson has suggested that the state space for population genetics would include the physically possible states of populations in terms of genotype frequencies. The state space would be "a Cartesian-space where '*n*' is a function of the number of possible pairs of alleles in the population" (Thompson 1983, p. 223). We can picture this geometrically as *n* axes, the values of which are frequencies of the genotype pair. The state variables are the frequencies for each genotype. Note that this is a one-locus system. That is, we take only a single gene locus, and determine the dimensionality of the model as a function of the number of alleles at that single locus.

Another type of single-locus system, used less commonly than the one described by Thompson, involves using single gene frequencies, rather than genotype frequencies, as state variables. Debates about "genic selectionism" center around the adequacy of this state space for representing evolutionary phenomena (see Sober and Lewontin 1982, for discussion of this issue). With both genotype and gene frequency state spaces, though, treating the genetic system of an organism as being able to be isolated (meaningfully) into single loci involves a number of assumptions about the system as a whole. For instance, if the relative fitnesses of the genotypes at a locus are dependent upon *other* loci, then the frequencies of a single locus observed in isolation will *not* be sufficient to determine the actual genotype frequencies. Assumptions about the structure of the system as a whole can thus be incorporated into the state space in order to reduce its dimensionality. Lewontin (1974) offers a detailed analysis of the quantitative effects on dimensionality of various assumptions about the biological system being modeled. It is made clear in his discussion that, although a state space incorporating the most realistic assumptions is desirable from a descriptive point of view, it is mathematically and theoretically intractable. For instance, a total genetic description (with no implicit assumptions) of a population with only two alleles at two loci would have a dimensionality of nine, whereas three alleles at three loci would be described in a 336-dimensional space (1974, p. 283). Most organisms have thousands of loci; the one-locus system is much more manageable, for example, for the formulation of laws of succession for the system.

A number of objections to the single-locus system have been raised by biologists. These objections, reviewed later, can be understood in terms of the descriptive inadequacy of the dimensions of the state space.

Michael Wade, in his discussion of group selection models, objects to the use of the single-locus model in calculations of the strength of group selection versus individual selection. Because some of the processes important to the operation of group selection (e.g., genotype-genotype interaction and interactions between loci) *cannot* be represented by a single-locus model, results of comparison of the forces of individual selection versus group selection within the context of such models is inevitably skewed (Wade 1978, pp. 103–104).

Interactions between genotypes, and between one locus and another, cannot (except in the case of frequency-dependent selection) be represented in a single-locus model, for the simple reason that they involve more than one locus. The trajectory of the frequency of a gene involved in these processes in a single-locus model will not follow a lawlike pattern, and will be thus inexplicable. Lewontin offers an example involving two polymorphic inversion systems whose frequencies are dependent on one another. The actual frequencies are inexplicable in a one-locus model, which does not allow for the interaction of the two polymorphisms in their determination of fitness. Models of higher dimensionality (or using different state variables) are necessary, because of the "dimensional insufficiency" of the single-locus models (1974, pp. 273–281; because this example has been discussed at length by Wimsatt 1980, pp. 226–229, I shall not go into detail here).

At this point, I would like to introduce an additional category. Although all single-locus models should, in some sense, be grouped together, they are not all exactly the same model – each particular model has a different number of state variables, depending upon the number of alleles at that locus. Van Fraassen has suggested calling the general outline for each model its "model type" (1980, p. 44). Because a model type is simply an abstraction of a model, constructed by abstracting one or more of the model's parameters, a single model can be an instance of more than one model type; the model types themselves are therefore not hierarchically arranged. Along similar lines, I suggest that each model type be associated with a distinctive *state space type*. In the preceding example, the single-locus model is to be taken as an instance of a general state space type for all single-locus models, that is, the different single-locus model types are conceived as using the same state space type. Alternatives, such as two-locus models (see Lewontin 1970 for an example), must be taken as instances of a different state space type.

Lewontin, dissatisfied with the theoretical results afforded by the use of single-locus and even multilocus state space types, has suggested an entirely different state space type. The intention is to treat the entire genome as a whole, rather than as a collection of independently segregating, noninteracting genotypes of single loci. Ernst Mayr has stressed the importance of the interaction of genes and the homeostasis of genotypes (i.e., large amount of linkage) in evolutionary processes. The genome will respond to selection pressures *as a whole*, says Mayr, instead of as an aggregate of individual loci (Mayr 1967, p. 53). In our terms, because evolution works this way, any accurate model of evolution cannot employ the single-locus state space type. Following up on his claim that the construction of a dynamically sufficient theory of a genome with many genes is "the most pressing problem of theory," Lewontin suggests an alternative approach using a completely different set of state variables (1974, p. 318).

According to the semantic view, a description of a theory's structure involves the description of the family of models for the theory. An essential part of this description of the family of models consists in describing the specific types of state spaces in terms of which the models are given. In this section, I have presented a general sketch of the types of state spaces associated with various model types, that is, a description of the class of state space types.

3.2. Parameters

Values that appear in the succession and coexistence laws of a system that are the same for all possible states of the defined system are here called *parameters*. For instance, in the modification of the Hardy-Weinberg equation that predicts the frequencies of the genotypes after selection, the selection coefficient, s, appears as a parameter in the equation:

$$p^2AA + 2pqAA + (1 - s)q^2aa = 1 - sq^2$$

There are a variety of methods of establishing the value at which a parameter should be fixed or set in the construction of models for a given real system. Simulation techniques, such as those presented in Section 2.3, can be used to obtain estimates of biologically important parameters. In some contexts, maximum likelihood estimations may be possible.

Parameters also can be set arbitrarily, or ignored. This is equivalent to incorporating certain assumptions into the model for purposes of simplification (see Section 3.1 on state space assumptions). (See also Levins 1968, pp. 8, 89; Bailey 1967, pp. 42, 220; Suppes 1967, pp. 62–63.)

Parameters can play roles of varying importance in the determination of the system represented by the theory. In this section, I shall discuss cases of the differing effect of the *values* of the parameters on the model outcome. The *choice* of parameters itself also can be theoretically important, as seen in the group selection example later.

One expects the values of parameters to have impact on the system being represented; but variations in parameter values can make a larger or smaller amount of difference to the system. For instance, take the deterministic model that incorporated a parameter for mutation, μ (see Section 2.1). The outcomes of this model are virtually insensitive to variations in the value of μ. Yet the selection parameter plays a crucial role in this same model; a very small amount of selection in favor of an allele will have a cumulative effect strong enough to replace other alleles (Lewontin 1974, p. 267).

Population size is another case in which the value assigned to the parameter has a large impact on the model results. As the case of polymorphism in the house mouse shows (discussed at length in Section 2.3), effective population size, *N*, can play a crucial role in some models, because selection results can be quite different with a restricted gene pool size (see Mayr 1967, pp. 48–50). In many of the stochastic models involved in calculating rates of evolutionary change, the resulting distributions and their moments can depend completely on the ratios of the mean deterministic force to the variance arising from random processes (Lewontin 1974, p. 268). This variance is usually proportional to *1/N*, and is related to the finiteness of population size. Thus, change in the value of the single parameter, *N*, can completely alter the structures represented by the theory.

The choice of parameters also can make a major difference to the model outcome. Theoreticians have choices about how to express certain aspects of the system or environment. The choice of parameters used to represent the various aspects can have a profound effect on the structure, even to the point of rendering the model useless for representing the empirical system in question. Group selection models provide a case in which choice of parameters not only altered the results of the models but also led to the near disappearance (in the models) of the phenomenon being modeled, according to Wade (see 1978).

Wade analyzed the major group selection models and found that they contain a number of common assumptions about various ecological processes, including extinction rates, migration, dispersion, and colonization (1978, pp. 103, 112). He challenged the accuracy of several of these assumptions on grounds that they are not biologically realistic enough. Take the assumption about colonization in the models – there are different *modes* of colonization, and the presence of these different modes has different effects on a number of factors affecting the existence and strength of group selection (1978, p. 103). Wade claims that the assumptions about colonization incorporated into existing models limit, automatically, several mechanisms for creating and maintaining genetic variation between populations (1978, p. 105; 1977, p. 150).

Because variation of group traits between groups is the analog in group level models to genetic variation in the genotype level models, and the operation of selection is dependent on variation among the units of selection, the initial assumptions about colonization can have large effects on the selection results of the system modeled. Existing assumptions regarding colonization are not representative, Wade claims, and he challenges them empirically. He finally suggests an alternate model incorporating new assumptions about colonization and new values for the colonization parameters suggested by his empirical research (1978, pp. 109–110; 1976; see Wimsatt's discussion, 1980, pp. 238–248).

Some authors, when discussing genetical changes in populations, speak of the system in terms of a phenotype state space type (Eden 1967, p. 10; Lewontin 1974, pp. 9–13). This makes sense, because the phenotype determines the breeding system and the action of natural selection, the results of which are reflected in *some* way, in the genetic changes in the population. In his analysis of the present structure of population genetics theory, Lewontin traces a single calculation of a change in genetic state through both genotypic and phenotypic descriptions of the population. That is, according to Lewontin, population genetics theory must map the set of genotypes onto the set of phenotypes, give transformations in the phenotype space, and then map the set of phenotypes back onto the set of genotypes. We would expect, then, that descriptions of state in population genetics would be framed in terms of *both* genetic and phenotypic variables and parameters. But this is not the case – the description can be in terms of *either* genotype *or* phenotype variables, but not both. Dynamically, then, it seems as if population genetics must operate in two parallel systems: one in genotype state space; one in phenotype state space (Lewontin 1974, pp. 12–13).

Lewontin explains that such independence of systems is illusory, "and arises from a bit of sleight-of-hand in which phenotype and genotype variables are made to appear as merely parameters that need to be experimentally determined, constants that are not themselves transformed by the evolutionary process" (1974, p. 15). A prime example of such a "pseudo parameter" is the fitness value associated with the individual genotypes while computing the mean fitness value, *w* bar. The mean fitness value appears in the equation which expresses the relative change in allele frequency, Δq, of an allele at a locus after one generation, in terms of the present allele frequency, q, and the mean fitness, *w* bar, of the genotypes in the population:

$$\Delta_q = \frac{q^{(1-q)}}{2} \frac{dln\,\bar{w}}{dq}$$

(Lewontin 1974, p. 13). Although *w* bar is used in computation in a genotype state space type, fitness is a function of phenotype, not genotype. Thus, information regarding values of phenotype variables is smuggled into the genotype models through parameters.[4]

3.3. Laws

In line with my goal of providing a general approach for describing the models of population genetics (because the theory is being described in terms of a family of models), I discuss in this section a few particular aspects and forms of the laws used in these models. The most obvious differences, in laws as well as in state space types, are between deterministic and stochastic models. But, as discussed later, even laws having the common framework of the Hardy-Weinberg equilibrium can differ fundamentally.

Coexistence laws describe the possible states of the system in terms of the state space. In the case of evolutionary biology, these laws would consist of conditions delineating a subset of the state space which contains only the biologically possible states. Changes in the state of the system are described by laws of succession. In the case of evolutionary theory,

[4] A referee for this journal points out that the situation is even more complicated than this passage suggests. That is, this phenotype information amounts to "the average effect of the phenotypes in fact produced by the relevant genotype in the present generation"; but changes elsewhere in the genome or any other changes in the genetic environment may yield a different "average" phenotype for the same genotype. Furthermore, the same average phenotype could yield very different fitnesses in slightly different environments.

dynamical laws concern changes in the genetic composition of populations[5] (see Lewontin 1974, pp. 6–19).

The laws of succession select the biologically possible trajectories in the state space, with states at particular times being represented by points in the state space (this is simplified – see discussion on time variables later). The law of succession is the equation of which the biologically possible trajectories are the solutions (van Fraassen 1970, pp. 330–331).

The Hardy-Weinberg equation, of which several variations were presented in Section 2, is the fundamental law of both coexistence and succession in population genetics theory. As Lewontin has noted, even the dynamical laws of the theory appeal to only the equilibrium states and steady-state distributions, which are estimated from the Hardy-Weinberg equation or variations thereof (1974, p. 269). The Hardy-Weinberg law is a very simple, deterministic succession law, which is used in a very simple state space. As parameters are added to the equation, we get *different* laws, technically speaking. For example, compare the laws used to calculate the frequency p' of the A allele in the next generation. Including only the selection coefficient into the basic Hardy-Weinberg law, we get $p' = p/(1 - sq^2)$. Addition of a parameter for mutation rate yields a completely different law, $p' = p - ps + \mu q$. We could consider these laws to be of a single type – variations on the basic Hardy-Weinberg law – which are usually used in a certain state space type. The actual state space used in each instance depends on the genetic characteristics of the system, and not usually on the parameters. For instance, the succession of a system at Hardy-Weinberg equilibrium and one which is *not* at equilibrium but is under selection pressure could both be modeled in the same state space, using different laws.

In the discussion in Section 2 involving equilibrium and dynamical models employing the Hardy-Weinberg equilibrium, the distinction between stochastic and deterministic models loomed large. Examination of the general features of the deterministic and statistical laws that appear in these models should help clarify the structure of the theory itself.

[5] The concept of a system changing over time, where the system is usually interpreted as a single population or species, is peculiar. David Hull has suggested that a more appropriate interpretation of such systems would be as *lineages*, which have the desirable qualities of being necessarily spatiotemporally localized and continuous (personal communication). Note, however, that such interpretive problems are a separate issue from description of the models, which simply represent ideal systems. Clearer understanding of the ideal systems and their interrelations should shed light on the advantages and disadvantages of the various possible empirical interpretations of the systems (e.g., see Hull 1980).

A theory can have either deterministic or statistical laws for its state transitions. Furthermore, the states themselves can be either statistical or nonstatistical. In population genetics models, gene frequencies often appear in the set of state variables; thus, the states themselves are statistical entities.

In general, according to the semantic view, a law is deterministic if, when all of the parameters and variables are specified, the succeeding states are uniquely determined (this definition of determinism and its advantages over other definitions are discussed in detail by van Fraassen 1972, pp. 306–321). In population genetics, this means that the initial population and parameters are all that is needed to get an exact prediction of the new population state (Lewontin 1967, p. 87).

Statistical laws are constructed by specifying a probability measure on the state space. The example presented in Section 2.2 entailed assigning probabilities (frequencies) to each distinct possible value of gene frequency. Thus, the probability measure is constructed by taking a certain value for the gene frequency, obtaining the joint distribution (in this case, through simulation), and making a new state space of probabilities on the old state space of gene frequencies.

Sometimes it is possible, depending on the variables and parameters in the laws, to translate a stochastic law on determinate states into a deterministic law on statistical states (van Fraassen 1970, pp. 333–334). In the case of population genetics models containing statistical states, this particular translation may not be possible, and the laws might remain statistical laws on statistical state variables. Consider, for example, the case of the polymorphism discussed in Section 2.3. The stochastic model actually contained more *relevant* information, (i.e., about population size) but less information in general, because it did not yield determinate values. Stochastic and deterministic models can thus contain more or less information, depending on the question being asked, and the aspects of the system or environment being included.

In the last part of this section, I would like to discuss briefly a related problem regarding the flexibility available in representing a given system.

In the representation of a system, a state can be conceived as a function of time, or not. That is, the state vector itself can be a function of time; the state is represented as a point, whereas the history of the system can be represented as a curve. On an alternative approach, the operator representing the magnitude can be a function of time; the history of the system would be represented as points in this state space, the different points representing different "possible worlds" or world histories (van Fraassen 1970, pp. 329–335).

Lewontin is interested in the biological usefulness of each of these possible ways to represent systems. He claims that although the usual mode of presentation is done (in our terms) by employing an instantaneous state space, the information presented thereby is not very interesting to the biologist. A description of the "time ensemble of states of a given population" would be much more useful, he claims (1967, p. 82). We might interpret this as a claim that a "possible worlds" representation would represent the information in a more useful way. But Lewontin seems to be saying more than this.

The case he is considering involves the following problem: In one case, the gene frequency, Q, of a certain allele is calculated using a series of randomly fluctuating, uniformly distributed values of the selection coefficient. In the other case, the same procedure is performed using the exact same set of selection coefficient values, except in reverse temporal order. The resulting values of Q are *different* for the two cases. In other words, in general, if the curves representing the paths of the selection coefficients of each population through time are not identical, *even though* they have the same mean, variance, and any other statistical measurement, the model outcomes will *not* necessarily be identical, because of the difference in temporal order of the values (Lewontin 1967, p. 84). Thus, if a possible worlds representation were possible, it would seem to contain more information about the system, because the time histories are preserved in a certain sense. If this is so, then there would probably be problems translating between the two possible types of system, that is, possible worlds and instantaneous state space (analogous to Heisenberg and Schrodinger pictures, respectively, in quantum mechanics). Are biological systems different from physical systems in that the descriptions of the systems, conceived as both a function of time and independent of time, are *not* both represented as two aspects of the same system in a Cartesian space? Lewontin explicitly claims that the gene frequencies of populations do not follow the law of large numbers (1967, p. 84). In any case, this poses an intriguing problem for future foundational research.

3.4. Interrelation of Models

The issue of the exact interrelations among the different model types of population genetics is the topic of another entire paper. Here I wish to make a few preliminary remarks.

According to the semantic view, the structure of a theory can be understood by examining the family of models it presents. In the case of population genetics theory, the set of model types – stochastic and deterministic, single-locus or multilocus – can be understood as a related family of models. The question then becomes defining the exact nature of the relationships among them.

One rather nice example of a detailed analysis of a relation among models was discussed in Section 3.2. There, parameters of genotype fitness were found to be versions of information about phenotypes, condensed into genetic form. The model types constructed on phenotype and genotype state space types can thus be understood as overlapping through the specific parameter of fitness.

It also can be useful to examine models of the same phenomenon that have different degrees of complexity. Some loss of information occurs in all models when the parameters are set. By fixing the value of or ignoring a factor that is known to be important in some contexts, assumptions are made that simplify the model.

Sometimes the incorporation of simplifying assumptions reduces the usefulness of the model. In the example presented in Section 2.3, the assumption, present in the deterministic model, that the effective population size had no bearing on the outcome of the model, turned out to render the model inferior to a model which omitted such an assumption. Lewontin and Dunn concluded that the latter model "more nearly explains what is observed in nature" since it is "closer to the real situation" (1960, p. 707).

It might seem that the inclusion in a model or set of models of assumptions that obviously do not correspond to the observed phenomena would necessarily detract from the usefulness or accuracy of the models involved. But the biologist Richard Levins has suggested a method of eliminating detrimental effects of arbitrary assumptions on a theory as a whole. He recommends replacing the unrealistic assumptions in a given model or model type with other (perhaps equally unrealistic) assumptions. A theorem that is supported by means of different models "having in common the aspects of reality under study but differing in other details, is called a robust theorem" (1968, p. 7). The actual operation and usefulness of such theorem testing is in the realm of theory confirmation, and will not be discussed here. The important point is that, through construction and comparison of alternative models and model types of a given phenomenon for purposes of confirmation, the interrelation among the various model types and parameters is made explicit.

4. CONCLUSION

In this paper, I have taken the semantic view of theory structure as a general framework for foundational studies of population genetics. A brief review of the semantic approach provided a basis on which to discuss the theory. I examined specific problem areas in this particular program of foundational studies, using examples of population genetics models presented in Section 2.

According to the semantic view, to present a theory is to present a related family of models. In the simplest case, these models are related by having a common state space and common law of succession. Differences between the models lie in different initial conditions (i.e., initial location of the system in the state space) and, hence, different successions of states satisfying the same law. Generally, however, the theory allows for greater essential variety in the system it deals with (for instance, different degrees of freedom). Representation of these systems then requires models with different state spaces (for instance, of different dimensionality) and different laws.

In Section 2, I reviewed a variety of models employed in population genetics. Using these as a starting point, in Section 3 I approached the task of describing the models of population genetics in terms of the classification of state space types (such as single-locus state spaces), and law types (such as the general form of the Hardy-Weinberg laws). In general, instances of a type differ by having different values for parameters, while different types result from the choice of parameters. Further research is indicated, especially on the relations between state spaces (such as in genotype and phenotype modeling) and the relations between different sorts of laws of succession (such as in deterministic and stochastic modeling).

3

Confirmation of Ecological and Evolutionary Models

1. INTRODUCTION

This paper concerns hypothesis testing and confirmation in evolutionary and ecological theory. I outline specific criteria used in evaluating evidence for theories, and demonstrate the use of each criterion through examples from various branches of evolutionary biology and ecology. The philosophical discussions in Roughgarden (1983), Strong (1983), Simberloff (1983), and Quinn and Dunham (1983), which focus on a Popperian approach to theory testing and acceptance, present some important issues in the testing of evolutionary and ecological explanations. I find that imprecision of criteria of testing and confirmation is the weakest point in these discussions. As an alternative to a Popperian approach (as defended by, e.g., Simberloff 1983), and to other approaches commonly cited by biologists (e.g., J. Platt's "strong inference"), I suggest a new description of confirmation that includes a detailed classification of the ways in which a theory may be confirmed (see, e.g., Oster and Wilson 1978 for an endorsement of Platt's 1964 paper).

Roughgarden (1983) proposes that one establishes an empirical fact in science "by building a convincing case for that fact." What counts as a convincing case depends, according to Roughgarden, on "common sense and experience" (pp. 583–584).

As Strong (1983) rightly points out, the appeal to common sense is problematic; common sense may not be "common" to all scientists concerned, and it says nothing about testing. Strong's solution to problems of confirmation and testing is not much of an improvement, though: "our

Lloyd, E. A. (1987). Confirmation of Ecological and Evolutionary Models. *Biology and Philosophy*, *2*, 277–293.

regard would be greatest for theories that have passed multiple independent, tough tests" (1983, p. 638). One is immediately inclined to ask what a "tough" test is, what constitutes "passing," and whether theories that have passed some test, but not "multiple, independent, tough tests," could be acceptable. Quinn and Dunham, in addressing the same problem, conclude, "Theories are embraced when, in part, a relatively simple explanation seems to account satisfactorily for much of a complex set of observations, and are abandoned or modified as the weight of post hoc additions becomes a burden, and other, comparably simple and appealing viewpoints are suggested" (1983, p. 613).

The chief weakness of these discussions is the lack of precision. It is unclear what it is for an explanation "to count satisfactorily" for a set of observations, or for a theory to "pass multiple, independent, tough tests," or for a set of evidence to count as "a convincing case." In what follows, I offer a view of confirmation that reveals various factors determining the support of theories by data. In doing so, I pursue a naturalistic approach to the philosophy of science, compatible with the approach recently defended by Ronald Giere (1985). I present a description of the various ways in which empirical claims about models can be confirmed; it is not assumed that all accepted models are supported in all of these possible ways. I would also like to emphasize that this schema is not intended as straightforwardly normative, that is, as a checklist for "good" or "well-confirmed" theories, but, rather, as a list of the types of support deemed significant within the disciplines of evolutionary biology and ecology. After a brief introduction to the various forms of confirmation, I discuss examples taken from evolutionary theory and ecology.

2. CONFIRMATION

Throughout this paper, I use the semantic view of scientific theories. On the semantic view, scientists present descriptions of ideal systems; a set of (logical) models can be defined relative to these structures. In presenting a scientific theory, scientists can be understood to be presenting sets of models, to be used to explain the world (for a detailed presentation of the semantic view, see Suppes 1957, 1967; van Fraassen 1970, 1972, 1980; Suppes 1972, 1977; for application of the semantic approach to evolutionary theory, see Beatty 1980, 1981, 1982; Thompson 1983, 1985; Lloyd 1983, 1984, 1986a, 1986b).

According to the state-space version of the semantic view (see van Fraassen, Suppe) models are specified by defining the variables with which the system in nature is described, the laws that describe the changes or structure of the system, and parameters, quantities in the model with a constant value.

Explanations involve claims regarding the applicability of a model to a natural system. Giere has characterized these claims, which he calls "theoretical hypotheses," as having the following general form: "The designated real system is similar to the proposed model in specified respects and to specified degrees" (1985, p. 80). For instance, a population geneticist might claim that a certain natural or laboratory population is a Mendelian system, that is, that it conforms with a Mendelian theoretical model. Certain attributes of the system – the distribution of gene frequencies, for instance – are thus explained through the homomorphism of the natural population to the theoretical model.

The activity of confirming models is more accurately described as confirming the empirical claims made about models, that is, the claims stating that a natural system (or kind of natural system) is homomorphic in certain respects to the model.

Patrick Suppes has given an analysis of the hierarchy of theories needed to link the natural system to the ideal system described by the theory (1962). In his (admittedly preliminary) study, Suppes presents three levels of models used to relate the empirical data to the theoretical model: theoretical models, models of the experiment, and models of data. The basic idea is that the logical models of the theory are too broad, because a model of the experimental data might fail to match precisely a theoretical model; for example, concepts or entities might be used in the theory that have no observable analogue in the experimental data (see Suppes 1962, p. 253). A series of gradually more specified models may be defined in order to make direct comparison possible.

The model of the experiment is the first step in specifying the theoretical model enough to enable comparison with the empirical results. The model of the experiment is a definition of all possible outcomes of a particular experiment which would satisfy the theoretical model. The next step in Suppes's hierarchy is the model of the data. In the context of a specific performance of an experiment, a portion of the total possible space of outcomes can be defined, each of which is a possible realization of the data. A possible realization of the data counts as a model of the data when it fits the model of the experiment well enough according to statistical goodness-of-fit tests. Models of data are usually restricted to those aspects of the experiment that have variables in the theory (1962, p. 258)

My use of "empirical claim regarding the model" is essentially equivalent to Suppes's "model of the experiment," in that it is more specified, more concrete than the abstract theory itself. For our purposes, the issues regarding "fit" (Sections 2.1, 2.3) can be understood in terms of statistical tests involving models of data and models of the experiment (in Suppes's terms), although I shall not use the distinction between models of the theory and models of the experiment. The issues involving independent support for aspects of the model (Sections 2.2, 2.3) remain outside the goodness-of-fit relations discussed by Suppes. Hence, I shall not use Suppes's terminology, although I see my description of confirmation as compatible with his.

In general, empirical evidence confirms a claim if the evidence gives additional reason to accept the claim. Evaluation of confirmation involves an evaluation of the support of claims regarding the applicability of a model to evidence, that is, an evaluation of the relation between the data and the model. Past discussions of confirmation have often been too general and too vague to be of real use. I take it that the semantic approach provides a precise framework within which theoretical, methodological, and empirical issues can be discussed.

I suggest that three distinct factors bear on the confirmation of empirical claims about models through data: fit between model and data; independent testing of aspects of the model; and variety of evidence, which can itself be of three sorts (see below). Traditionally, philosophical discussions of confirmation have concentrated on the fit between data and model, and on variety of instances of fit, which is one type of variety of evidence. Detailed examples of each type of confirmation will be given after a brief summary of the types.

2.1. Fit between Model and Data

The most obvious way to support a claim of the form "this natural system is described by the model," is to demonstrate the simple matching of some part of the model with some part of the natural system being described. For instance, in a population genetics model, the solution of an equation might yield a single genotype frequency value. The genotype frequency is a variable in the model. Given a certain set of input variables (e.g., the initial genotype frequency value, in this case), the output values of variables can be calculated using the rules or laws of the model. The output set of variables (i.e., the solution of the model equation given the input values of the variables) is the outcome of the model. Determining the fit involves testing how well the genotype frequency value calculated from the model (the outcome of the model) matches the genotype frequency measured in relevant natural populations. Fit can be evaluated by determining the fit of one curve (the model trajectory or coexistence conditions) to another (taken from the natural system); ordinary statistical techniques of evaluating curve-fitting are used for this evaluation.

2.2. *Independent Support for Assumptions*

Numerous assumptions are made in the construction of any ideal system. These include assumptions about which factors influence the changes in the system, what the ranges for the parameters are, and what the mathematical form of the laws is. On the basis of these assumptions, the models are given certain features. Many of these assumptions have potential empirical content. That is, although they are assumptions made about certain mathematical entities and processes during the construction of the ideal system, when empirical claims are then made about this ideal system, the assumptions may have empirical significance. For instance, the assumption might be made during the construction of a model that the population is panmictic, that is, that all genotypes interbreed at random with each other. The model outcome, in this case, is still a genotype frequency, for which ordinary curve-fitting tests can be performed on the natural population to which the model is applied. But the model can have additional empirical significance, given the empirical claim that a natural system is a system of the kind described in the model. The assumption of panmixia, as a description of the population structure of the system under question, must be considered part of the system description that is being evaluated empirically. Evidence to the effect that certain genotypes in the population breed exclusively with each other (i.e., evidence that the population is far from panmictic) would undermine empirical claims about the model as a whole, other things being equal. In other words, the assumption that genotypes are randomly redistributed in each generation is intrinsic to the Hardy-Weinberg equilibrium. Hence, although the assumption that the population is panmictic appears nowhere in the actual definition of the model type – that is, in the law formula – it is interpreted empirically, and plays an important role in determining the empirical adequacy of the claim.

By the same token, evidence that the assumptions of the model hold for the natural system being considered will increase the credibility of the claim that the model accurately describes the natural system.

In other words, it is taken that direct testing provides a stronger test than indirect testing, hence a higher degree of confirmation if the test is supported by empirical evidence. Direct empirical evidence for certain empirically interpreted aspects of the model that are not included in the state variables (and thus are confirmed only indirectly by goodness-of-fit tests) therefore provides additional support for the application of the model.

This sort of testing of assumptions involves making sure that the empirical conditions for application of the ideal description actually do hold. In order to accept an explanation constructed by applying the model, the conditions for application must be verified.

The specific values inserted as the parameters or fixed values of the model are another important aspect of empirical claims. In some models, mutation rates, and so on, appear in the equation – part of the task of confirming the application of the model involves making sure that the values inserted for the parameters are appropriate for the natural system being described.

Finally, there is a more abstract form of support available, in which some general aspect of the model, for instance, the interrelation between the two variables, or the significance of a particular parameter or variable, can be supported through evidence outside the application of the model itself.

2.3. Variety of Evidence

Variety of evidence, of which there are three kinds, is an important factor in the evaluation of empirical claims. I discuss three kinds of variety of evidence here: (1) variety of instances of fit; (2) variety of independently supported model assumptions; and (3) variety of types of support, which include fit and independent support of aspects of the model.

Often, empirical claims are made to the effect that a model is applicable over a more extended range than that actually covered by available evidence. This extrapolation of the range of a model can be performed by simply accepting or assuming the applicability of the model to the entire range in question. A more convincing way to extend applicability is to offer evidence of fit between the model and the data in the new part of the range. Provision of a variety of fit can thus provide additional reason for accepting the empirical claim regarding the range of applicability of the model. For instance, a theory confirmed by ten instances of fit involving populations of size 1000 (where population size is a relevant parameter) is in a different situation with regard to confirmation than a theory confirmed by one instance of each of ten different population sizes ranging from 1 to 1,000,000. If the empirical claims made about these two models asserted the same broad range of applicability, the latter model is confirmed by a greater variety of instances of fit. That is, the empirical claim about the latter model is better confirmed, through successful applications (fits) over a larger section of the relevant range than the first model. Variety of instances of fit can therefore provide additional reason for accepting an empirical claim about the range of applicability of a model.

Variety of fit is only one kind of variety of evidence. An increase in the number and kind of assumptions tested independently, that is, greater variety of assumptions tested, also would provide additional reason for accepting an empirical claim about a model. This is just the sort of confirmation Thompson (1985) found lacking in many sociobiological explanations of human behavior. Thompson argues that the difference in the acceptability of sociobiological explanations of insect behavior and of human behavior is that the auxiliary theories used in applying genetic models to human beings are unsupported. For instance, there is no neat physiological model linking genes to a particular phenotype in the case of homosexuality, as there is, for example, in sickle cell anemia (see Thompson 1985, pp. 205–211). The assumptions needed to apply the genetic models to human beings are largely unsupported empirically, and on these grounds, the sociobiological explanations are rejected, argues Thompson.

The final sort of variety of evidence involves the mixture of instances of fit and instances of independently tested aspects of the model. In this case, the variety of types of evidence offered for an empirical claim about a model is an aspect of confirmation.

According to the view of confirmation sketched earlier, claims about models may be confirmed in three different ways: (1) through fit of the outcome of the model to a natural system; (2) through independent testing of assumptions of the model, including parameters and parameter ranges; (3) through a range of instances of fit over a variety of the natural systems to be accounted for by the model, through a variety of assumptions tested, and through a variety of types of evidence.

3. APPLICATIONS

3.1. Fit

3.1.1. ISLAND BIOGEOGRAPHY The area of island biogeography offers a straightforward example of confirmation of a model application through its fit with empirical findings. The first attempt at a quantitative theory of island biogeography was made by MacArthur and Wilson (1963, 1967). They presented a mathematical model for determining the equilibrium numbers of species on islands. In this model, the equilibrium number of species is represented by the point of intersection of the immigration and extinction curves of the island, which are drawn as a function of the number of species already present and the distance from the mainland.

The assumptions of the model include speculations about the equations of the curves, and about the effects of varying both island size and the distance from the island to the source of the immigrating fauna.

In the first empirical tests of the MacArthur-Wilson model, the investigators assumed that it was most important to show: (1) that there exists an equilibrium number of species, and (2) that the MacArthur-Wilson model accurately represents the relationship among the species equilibrium number and the species turnover rate (see Wilson and Simberloff 1969).

Wilson and Simberloff chose to test the two empirical claims above by removing fauna from seven very small islands (Mangrove Islands in the Florida Keys), and surveying the colonization results. The results are taken to support the empirical claims outlined earlier.

In support of the claim that an equilibrium number of species does exist, Simberloff and Wilson (1969) cite three types of evidence: first, the number of species on the control (nondefaunated) islands did change during the period of the experiment; second, untreated islands with similar area and distance from source faunas have similar equilibrium numbers of species to those arrived at on the experimental islands; and, third, there was an increase of species on the experimental islands to approximately the same number as before defaunation, and then oscillation around this number.

These types of evidence can be taken as instances of fit between the outcome of the model (the equilibrium number of species) and the empirical findings (the actual number of species, and the pattern over time). Later, independent investigations found additional support from other bodies of data for the existence of an equilibrium number of species (Diamond, 1969).

Simberloff and Wilson also examined the actual turnover rate on the islands, and compared it with the predictions resulting from the MacArthur-Wilson model. They found that the experimental results were roughly consistent with the model prediction. Again, the outcome of the model (somewhat loosely) fitted the empirical findings.

In follow-up research on these islands, the investigators offer further confirmation for the existence of an equilibrium number of species and the accuracy of the model equation through further instances of fit (Simberloff and Wilson, 1970).

3.1.2. PUNCTUATED EQUILIBRIUM Let us examine the type of support offered for the controversial theory of punctuated equilibrium, presented by Niles Eldredge and Stephen Jay Gould (Eldredge 1971; Eldredge and Gould 1972; Gould and Eldredge 1977). For all the controversy, the theory of punctuated equilibria is a relatively simple sort of model, which has two main features. First, speciation by branching of lineages (as in allopatric speciation, i.e., speciation by geographic isolation, see Mayr 1963) is the primary source of significant evolutionary change, according to the model, rather than the gradual transformation of lineages (phyletic transformation). Second, speciation occurs rapidly in geologic time, and is followed by long periods of stasis.

As Gould emphasized, the model presents a picture of the relative frequencies of gradual phyletic transformation and punctuated equilibrium (Gould, 1982). Gould and Eldredge's 1977 paper contains a section entitled "Testing Punctuated Equilibria," in which they present various approaches to the fitting of the model to the data. Because punctuated equilibrium is a model about relative frequency, the general approach to confirmation is to test the distribution of instances. Gould and Eldredge discuss two ways in which the frequency distribution can be tested. First, the model can be applied to individual cases (of evolutionary change) with the right sort of features. The authors discuss the merits of a number of cases of individual fit. In some cases, Gould and Eldredge claim that the data presented as confirmation for the gradualist model actually have a tighter fit to their punctuated equilibria model (see especially the discussion of Gingerich 1977, pp. 130–134). In all of these cases, the issue is whether the data presented conform sufficiently to the predictions or structures of the models in question.

A second type of test involves examination of quantifiable features of entire clades or communities, and comparison of these features to results expected from the model. Stanley (1975) devised a number of this sort of high-level test for the punctuated equilibria model. In several tests, he demonstrates that, given the estimated time span and major morphological evolution, the gradualist model produces rates of evolutionary change that are too slow. Under the punctuated equilibria model, however, the evolutionary changes could conceivably have taken place within the estimated time span. The support being offered for the theory of punctuated equilibria here is that it has a better fit with the data than the gradualist model (see Stanley 1975, 1979; Gould and Eldredge 1977, pp. 120–121).

3.1.3. STATISTICAL POWER Statistical analysis is commonly used in evaluation of the fit of the empirical data to the model. A recent discussion on the power of statistical tests emphasizes that there is more to fit of model to data than maintaining a level of alpha (Type I error) under .05, where the probability of committing a Type I error is the probability of mistakenly rejecting a true null hypothesis (Toft and Shea, 1983). Toft and Shea argue that the probability of committing Type II errors (in which the investigator mistakenly fails to reject a false null hypothesis) is important, and has been neglected. The "power" of a test is the probability of not committing a Type II error.

A specific problem has arisen regarding power of tests in investigations of competition theory in ecology – the failure to demonstrate that a certain factor has an effect on a system is sometimes taken as the demonstration that the factor has no effect. Such a conclusion is unwarranted by the evidence, as well as by statistical theory, as Toft and Shea point out, and they call on ecological investigators to include power tests in their results in the future (Toft and Shea, 1983).

Toft and Shea's criticism of the investigators' method can be understood as an elaboration of the definition of a "good fit." It is not enough, they argue, to have a low probability of Type I errors in evaluating the fit of a model to a natural system; without consideration of Type II errors, tightness of fit is open to misinterpretation, and can be used to support false claims.

3.2. Independent Testing

3.2.1. PUNCTUATED EQUILIBRIUM To return to the topic of punctuated equilibria, Gould and Eldredge, in their discussion of the empirical support for the model, included a section on "indirect testing" (see 1977, pp. 137–129). Under the schema used in this paper, these tests can be understood as independent tests of the empirical assumptions of the model type.

The situation is as follows: one primary assumption of the model of punctuated equilibrium is that a major amount of genetic change can and does occur in the speciation event itself. That is, the major genetic differences between species must be laid down during the process of speciation rather than gradually through the duration of the species' existence. Gould and Eldredge consider evidence for the concentration of genetic change in speciation events as an important possible source of confirmation of their theory. For instance, in one case (although they found much of the evidence ambiguous with regard to their model), they reported, "We are pleased that some recent molecular evidence . . . supports our model" (1977, p. 138). The case in point offers evidence for the concentration of genetic change in speciation, and thus provides independent confirmation for an important assumption of their model.

Note the difference between fit and independent testing. In the case of fit, the model outcome, which involves the relative frequencies of gradual phyletic transformation versus punctuated equilibria, is compared to the actual frequencies of these forms of speciation. In independent testing, an assumption that is important in constructing the model is evaluated separately from the model itself.

3.2.2. POPULATION GENETICS Population genetics theory consists of a large number of related models based on the Hardy-Weinberg equilibrium, which is in turn based on Mendel's basic laws of inheritance. Because the Hardy-Weinberg "law" is an equilibrium equation, any mathematical descriptions of changes in the system being described must involve parameters inserted into expanded versions of the Hardy-Weinberg equation which produce the correct changes in the models. For instance, take a model that will give you the gene frequency of A in the next generation. A large amount of change mutation from a to A will effect the outcome of the model, so the mutation rate, μ, of a to A is included as a parameter in the model. Similarly with migration and selection. In other words, models for gene frequency changes must include factors that visibly affect gene frequencies.

Population genetics models, for the most part, yield single gene frequencies or distributions of gene frequencies. These frequencies and distributions are the part of the model tested for fit. But empirical and experimental testing is also done on the parameter values and their ranges. The subject of mutation and genetic variability, for instance, has served as a central issue in population genetics research. Although the theoretical problems will not even be mentioned here, the point is that it has been vital to the success and acceptance of population genetics models that the empirical assumptions and parameter values be tested (see Dobzhansky 1970; Lewontin 1974; Mayr 1982).

Determining the value of the mutation parameter of a particular gene is a task theoretically and methodologically distinct from testing the accuracy of the population genetics model in determining gene frequencies. For example, one can find tables of mutation rates for various genes in various organisms; these tables are the results of counting (traditionally done through inbreeding experiments), rather than being the results of calculations done using the Hardy-Weinberg equilibrium. The general idea is that the parameter is isolated and tested separately from the model in which it appears.

3.2.3. GROUP SELECTION Michael Wade, in his discussion concerning models of group selection, examines certain key assumptions common to the models. Wade argues that these (speculative) assumptions are unfavorable to group selection, that is, in these models, group selection is considered an important cause of changes in gene frequency only under a very narrow range of parameter values. Wade challenges the empirical adequacy of the model assumptions and suggests alternative assumptions derived from empirical results (Wade 1978).

The mathematical models of group selection examined by Wade involve general assumptions about extinction, dispersion, and colonization. One of the five assumptions challenged by Wade is that group selection and individual selection always operate in opposite directions. That is, it is assumed in the models that an allele that is favored by selection between groups would be selected against on an individual level. According to Williams (1966), group selection should never be called up if individual selection can be used to explain the evolution of a trait. Thus, if individual selection and group selection are in the same direction, group selection might never be appealed to. Group selection will only be discussed in cases in which it operates in opposition to individual selection.

Wade argues against this assumption, pointing out that any trait favored by group selection (i.e., any trait that increases the likelihood of successful proliferation of the population or decreases the likelihood of extinction) also could be favored by individual selection. He offers experimental results of group and individual selection acting in the same direction (see Wade 1976, 1977, 1978).

3.3. Variety of Evidence

3.3.1. DARWIN'S EVIDENCE Darwin considered the large variety of evidence supporting his theory of evolution by natural selection to be partial grounds for accepting the theory. He wrote, "I have always looked at the doctrine of natural selection as an hypothesis, which if it explained several large classes of facts, would deserve to be ranked as a theory deserving acceptance" (1903, 1:139–140).

As far as Darwin was concerned, the theory of natural selection does account successfully for "several large classes of fact," including the principle of classification of living things, embryonic resemblance among organisms of very different taxa, and the adaptation of living beings to each other and to their environments (1919, 2:207). In other words, models based on the concept of natural selection fit the empirical observations in a wide variety of fields. This is an example of the first kind of variety of evidence, that is, variety of fit. The existence of adaptive characters in organisms is accounted for by referring to a natural selection model in which those organisms that are well adapted to their environment survive and reproduce at a proportionally higher rate than organisms without the adaptive mechanisms. Eventually, then, the adaptive mechanism would be expected to become a fixed trait in the population (given the right conditions of heredity, etc.). Similarly, natural selection models can account, for instance, for the strange fact that at a certain age, it takes a trained eye to distinguish human from chicken from fish embryos (Darwin 1964, pp. 439–440). The resemblance is understood as a result of common ancestry. Common ancestry also holds the key to taxonomic classification; classification becomes genealogy – the tracing of lineages – according to Darwin (1919, 1:485; 1964).

Thus, Darwin took it as a virtue of his theory of natural selection that it could account for a wide range of natural phenomena, that is, that it exhibited variety of evidence (see Lloyd 1983, for discussion).

3.3.2. POPULATION GENETICS Empirical support for basic population genetics model types (e.g., single-locus models based on the Hardy-Weinberg equilibrium) exhibits a different sort of variety of evidence from that claimed by Darwin for the theory of natural selection. The mathematical models used to calculate equilibria and changes in gene frequencies have been shown to fit a wide range of natural populations (e.g., those summarized in Dobzhansky, 1970). In addition, the parameters of many population genetics models have been evaluated and tested separately, as discussed in Section 3.2.2 of this paper.

Hence, population genetics models are not confirmed by fit alone but also by independent testing of model assumptions and a pattern of fits over a range of actual populations. The empirical support for population genetics represents more than one type of evidence, that is, it exhibits the third kind of variety of evidence, namely, variety of types of support.

3.3.3. MARINE ECOLOGY Part of a recent upheaval in ocean ecology involves a debate about the appropriate range of testing for a model, that is, the variety of fit of the model being tested. Before 1979, it was thought that phytoplankton growth rates and nutrient uptake (usually ammonium uptake) rates were coupled temporally. Models constructed to represent these rates under steady-state conditions had been successful in laboratory studies. Assuming the accuracy of this model type, however, led to a puzzle about the natural, oceanic systems: studies of oceanic waters showed the level of nitrogenous nutrients (primarily ammonium and nitrate) to be undetectably low, even though the data from photosynthetic activity indicated that the phytoplankton was absorbing nitrogenous nutrients (McCarthy and Goldman 1979, p. 670).

Furthermore, research on the chemical composition of laboratory and oceanic phytoplankton showed that the chemical composition of oceanic populations was most similar to laboratory populations growing at near maximal rates. However, in order to grow at such high rates, phytoplankton need high ambient nutrient levels (Goldman, McCarthy, and Peavey 1979). Thus, their chemical composition suggested that oceanic phytoplankton must be experiencing high nutrient levels, but the open ocean nutrient levels were extremely low.

Some light was shed on the puzzle when laboratory studies showed that phytoplankton are capable of rapid nutrient uptake. This, and the ability to store nitrogen, would make it possible for them to have a maximum growth rate, even when nutrient concentrations are very low. If rapid nutrient uptake occurs, then it is not necessarily true that growth rates are tied to nutrient uptake rates, as assumed in the previously accepted models, and found in laboratory studies carried out at steady state (Goldman, McCarthy, and Peavey 1979, p. 213). McCarthy and Goldman speculated that individual phytoplankton cells might encounter minute zones of elevated nitrogen levels, which they could absorb very quickly (1979). This would account for high rates of production despite the low observed nutrient level.

Phenomena of such a small scale are not taken into account when considering steady-state situations. Under steady-state conditions, the medium is homogeneous in space and time. That is, organisms are not subject to a feast and famine existence; therefore, rapid nutrient uptake and storage are phenomenologically invisible. Goldman et al. write: "to explore further questions [concerning nutrient dynamics] involves new approaches for studying microbial interactions on temporal and spatial scales that are far smaller than were previously assumed to be important" (1979, p. 214).

Later studies confirmed that differences in methods of measuring parameter values made significant differences to the experimental results. It had usually been assumed that the rate of nutrient uptake was linear over the course of the hours or tens of hours in an experiment. When short-term nutrient uptake responses were tested, however, they were found to be nonlinear (Goldman, Taylor, and Glibert 1981). In one experiment, the phytoplankton completed uptake of the nitrogenous nutrients during the first two hours of the experiment. If the measurements had been performed in the usual time span, for example, after 24 hours, the estimates of nitrogen turnover rates would have been an order of magnitude off. Estimates of phytoplakton growth rate based on these nitrogen turnover rates would in turn have been "in gross error" (Goldman, Taylor, and Glibert 1981, p. 146). The investigators concluded that choice of incubation period can have serious consequences in hypothesis testing (Goldman, Taylor, and Glibert 1981, p. 137).

This situation can be redescribed in the terms of our confirmation schema. When tested against steady-state laboratory systems, the model linking phytoplankton nutrient uptake rate to growth rate seemed adequate. A problem arose, however, when open ocean systems were found to contain very low nutrient levels, whereas phytoplankton were apparently growing at maximal rates. Investigators later extended the range of experimental systems over which the model was tested. In this case, the time component of the system definition was expanded significantly, to include short-range tests. When tests spanning seconds to minutes were performed, the steady-state model was found not to fit the data. This has led to the suggestion by the biologists that models representing a phytoplankton system must be tested against an incubation period based on the "time scale of physiological responses by phytoplankton" (Goldman,

Taylor, and Glibert 1981, p. 137). Such experiments have led to suggestions of new models for oceanic nutrient dynamics, using these shorter time scales.

4. CONCLUSION

The taxonomy of confirmation presented in this paper included empirical support for models in the form of: (1) fit of the model to data; (2) independent testing of various aspects of the model; and (3) variety of evidence. I presented examples of each type of confirmation, drawing from a range of evolutionary and ecological theories. Instances in which scientists criticize other investigators for lack of sufficient support of a given type are included. I have not attempted to analyze or to justify the various forms of support – that is a project for another paper. Rather, I have attempted to establish the plausibility and importance in evolutionary biology of different categories of empirical support. The greater complexity and variety in my approach, as compared to, for example, a Popperian approach, can facilitate detailed analysis and comparison of empirical claims.

ACKNOWLEDGMENTS

Special thanks go to Richard Lewontin for pressing the issue of confirmation with me, to David Glaser for his suggestion for examples, and to Bas van Fraassen for his unrelenting support. I also would like to thank Marjorie Grene, Michael Bradie, Hamish Spencer, E. O. Wilson, and Peter Taylor for their comments, questions, criticism, and suggestions.

4

Units and Levels of Selection

1. INTRODUCTION

Richard Lewontin was the first to investigate systematically the set of problems raised by a hierarchical expansion of selection theory in his landmark 1970 article. His classic abstraction and analysis of the three principles of evolution by natural selection – phenotypic variation, differential fitness, and heritability of fitness – have served as the launching point for many biologists and philosophers who have wrestled with units of selection problems. Lewontin's critical discussion of the empirical evidence for selection at various biological levels has served as both touchstone and target for later work. But Lewontin's essay was addressed to the efficacy of different units of selection as causes of evolutionary change (1970, p. 7). In the analysis I offer in this chapter, this is but one of four distinct questions involved in the contemporary units of selection debates.

For at least two decades, some participants in the "units of selection" debates have argued that more than one question is at stake. Richard Dawkins, for instance, introduced the terms replicator and vehicle to stand for different roles in the evolutionary process (1978; 1982a; 1982b). He proceeded to argue that the units of selection debate should not be about vehicles, as it had formerly been, but about replicators. David Hull, in his influential article, "Individuality and Selection" (1980), suggested that Dawkins's "replicator" subsumes two distinct functional roles, and

Lloyd, E. A. (2001). Units and Levels of Selection. In R. Singh, C. Krimbas, D. Paul, and J. Beatty (Eds.), *Thinking About Evolution* (pp. 267–291). Cambridge: Cambridge University Press.

the separate categories of "replicator," "interactor," and "evolver" were born. Brandon, arguing that the force of Hull's distinction had not been appreciated, analyzed the units of selection controversies further, claiming that the question about interactors should more accurately be called the "levels of selection" debate to distinguish it by name from the dispute about replicators, which he allowed to keep the "units of selection" title (1982).[1]

My purpose in this chapter is to delineate further the various questions pursued by Robert Brandon, Richard Dawkins, James Griesemer, David Hull, Richard Lewontin, John Maynard Smith, Sandra Mitchell, Elliott Sober, Michael Wade, George C. Williams, David S. Wilson, William Wimsatt, Sewall Wright, and many others under the rubric of "units of selection."[2] I will isolate four quite distinct questions that have, in fact, been asked in the context of considering, What is a unit of selection? In Section 2, I describe each of these distinct questions. In Section 3, I return to the sites of several very confusing, occasionally heated debates about "the" units of selection. I analyze many leading positions on the issues using my taxonomy of questions.

This analysis does not, of course, make differences vanish, but I hope to clarify the terms of the debates. My analysis also does not resolve any of the conflicts about which research questions are most worth pursuing; moreover, I do not attempt to decide which of the questions or combinations of questions I discuss ought to be considered *the* units of selection question. Although I have elsewhere argued that the interactor question (see Section 2.1) is a primary question for evolutionary genetics, that claim is intended as historical and descriptive; most evolutionary genetics models that address any version of the units of selection question have focused on which level of interaction must be represented in the model to make it dynamically and empirically adequate (Lloyd 1988, especially Chapters 5 and 6).[3] Furthermore, the mere persistence of the three other questions to be discussed attests to their importance and general interest.

[1] Mitchell also argues for the importance of keeping the interactor issue separate from issues involving replicators (1987).

[2] For some of the pivotal arguments in the debates, see Brandon 1982, 1985, 1990; Dawkins 1978, 1982a,b, 1986; Hull 1980; Lewontin 1970; Maynard Smith 1964, 1976; Sober 1984; Sober and Lewontin 1982; Wade 1978, 1985; Wilson 1975, 1980; Wright 1980.

[3] I have found nearly two hundred references to papers and books by biologists and philosophers that treat what is called here "the interactor question" (Lloyd 1988/1994); this represents just a fraction of the literature on the topic. The second most prominent interpretation of "the units of selection" question is "the replicator question" discussed in Section 2.2.

2. FOUR BASIC QUESTIONS

Four basic questions can be delineated as distinct and separable. As we shall see in Section 3, these questions are often used in combination to represent *the* units of selection problem. But before we continue, we need to clarify some terms.

The term *replicator*, originally introduced by Dawkins but since modified by Hull, is used to refer to any entity of which copies are made. Dawkins classifies replicators using two orthogonal distinctions. A "germ-line" replicator, as distinct from a "dead-end" replicator, is "the potential ancestor of an indefinitely long line of descendant replicators" (1982a, p. 46). For instance, DNA in a chicken's egg is a germ-line replicator, whereas that in a chicken's liver is a dead-end replicator. An "active" replicator is "a replicator that has some causal influence on its own probability of being propagated," whereas a "passive" replicator is never transcribed and has no phenotypic expression whatsoever (1982a, p. 47). Dawkins is especially interested in *active germ-line replicators*, "since adaptations 'for' their preservation are expected to fill the world and to characterize living organisms" (1982a, p. 47).

Dawkins also introduced the term *vehicle*, which he defines as "any relatively discrete entity... which houses replicators, and which can be regarded as a machine programmed to preserve and propagate the replicators that ride inside it" (1982b, p. 295). According to Dawkins, most replicators' phenotypic effects are represented in vehicles, which are themselves the proximate targets of natural selection (1982a, p. 62).

Hull, in his introduction of the term *interactor*, observes that Dawkins' theory has replicators interacting with their environments in two distinct ways: they produce copies of themselves, *and* they influence their own survival and the survival of their copies through the production of secondary products that ultimately have phenotypic expression. Hull suggests the term *interactor* for entities that function in this second process. An *interactor* denotes that entity that interacts, as a cohesive whole, directly with its environment in such a way that replication is differential – in other words, an entity on which selection acts directly (Hull 1980, p. 318). The process of evolution by natural selection is "a process in which the differential extinction and proliferation of interactors cause the differential perpetuation of the replicators that produced them" (Hull 1980, p. 318; cf. Brandon 1982, pp. 317–318).

Hull also introduced the concept of "evolvers," which are the entities that evolve as a result of selection on interactors; these are usually what

Hull calls "lineages" (Hull 1980). So far, no one has directly claimed that evolvers are units of selection. They can be seen, however, to be playing a role in considering the questions of who owns an adaptation and who benefits from evolution by selection, which we will consider in Sections 3.1 and 3.3.

2.1. The Interactor Question

In its traditional guise, the interactor question is, What units are being actively selected in a process of natural selection? As such, this question is involved in the oldest forms of the units of selection debates (Darwin 1859 (1964); Haldane 1932; Wright 1945). In his classic review article, Lewontin's purpose was "to contrast the levels of selection, especially as regards their efficiency as causes of evolutionary change" (1970, p. 7). Similarly, Slobodkin and Rapoport assumed that a unit of selection is something that "responds to selective forces as a unit – whether or not this corresponds to a spatially localized deme, family, or population" (1974, p. 184).

Questions about interactors focus on the description of the selection process itself, that is, on the interaction between entity and environment and on how this interaction produces evolution; they do not focus on the outcome of this process (see Wade 1977; Vrba and Gould 1986). The interaction between some interactor and its environment is assumed to be mediated by "traits" that affect the interactor's expected survival and reproductive success. (NB An interactor may be at any level of biological organization, including a group, an organism, a chromosome, a kin group, or a gene.) In other words, some portion of the expected fitness of the interactor is directly correlated with the "value" of the trait in question. The expected fitness of the interactor is commonly expressed in terms of genotypic fitness parameters, that is, in terms of the fitness of combinations of replicators; hence, interactor success is most often reflected in, and counted through, replicator success. Several methods are available for expressing such a correlation between trait and (genotypic or organismic) fitness, including regression and variances, and covariances. Several models are also available for representing interactors; in all of these, the interactor's trait is correlated with replicator fitness values, and the component of the replicator fitnesses attributed to the interactor is not available or reproducible from a lower level of interactor.[4]

[4] For example, Arnold and Fristrup 1982; Colwell 1981; Craig 1982; Crow and Aoki 1982; Crow and Kimura 1970; Damuth and Heisler 1988; Hamilton 1975; Heisler and Damuth 1987; Lande and Arnold 1983; Li 1967; Ohta 1983; Price 1972; Uyenoyama 1979;

In fact, much of the interactor debate has been played out through the construction of mathematical genetical models. The point of building such models is to determine what kinds of selection, operating on which levels, may be effective in producing evolutionary change.

It is widely held, for instance, that the conditions under which group selection can effect evolutionary change are quite stringent and rare. Typically, group selection is seen to require small group size, low migration rate, and extinction of entire demes.[5] Some modelers, however, disagree that these stringent conditions are necessary. Matessi and Jayakar, for example, show that in the evolution of altruism by group selection, very small groups may not be necessary (1976, p. 384; contra Maynard Smith 1964); Wade and McCauley also argue that small effective deme size is not a necessary prerequisite to the operation of group selection (1980, p. 811). Similarly, Boorman shows that strong extinction pressure on demes is not necessary (1978, p. 1909). And finally, Uyenoyama (1979) develops a group selection model that violates all three of the "necessary" conditions usually cited.

That different researchers reach such disparate conclusions about the efficacy of group selection is partly because they are using different models with different parameter values. Wade (1978) highlighted several assumptions, routinely used in group selection models, that biased the results of these models against the efficacy of group selection. For example, he noted that many group selection models use a specific mechanism of migration; it is assumed that the migrating individuals mix completely, forming a "migrant pool" from which migrants are assigned to populations randomly. All populations are assumed to contribute migrants to a common pool from which colonists are drawn at random. Under this approach, which is used in all models of group selection before 1978, small sample size is needed to get a large genetic variance between populations (Wade 1978, p. 110).

If, in contrast, migration occurs by means of large propagules, higher heritability of traits and a more representative sampling of the parent population will result. Each propagule is made up of individuals derived from a single population, and there is no mixing of colonists from different populations during propagule formation. On the basis of Slatkin and

Uyenoyama and Feldman 1980; Wade 1978, 1980, 1985; Wade and Breden 1981; Wade and McCauley 1980; Wilson 1983; Wilson and Colwell 1981; Wimsatt 1980, 1981. See discussion in Lloyd 1988.

[5] See Aoki 1982; Boorman and Levitt 1973; Fisher 1930; Ghiselin 1974; Leigh 1977; Levin and Kilmer 1974; Maynard Smith 1964, 1976; Uyenoyama 1979; Williams 1966.

Wade's analysis, much more between-population genetic variance can be maintained with the propagtile model (1978, p. 3531). They conclude that, by using propagule pools as the assumption about colonization, one can greatly expand the set of parameter values for which group selection can be effective (Slatkin and Wade 1978, p. 3531; cf. Craig 1982).

My point here is not, however, to survey the various models[6] but rather to illustrate the *level of disagreements* within the interactor question itself. It should also be emphasized that not all discussion regarding which levels of selection are causally efficacious has been quantitative. Many authors have attempted to determine which levels of selection must or should be taken into account through qualitative descriptions of interactors.[7]

Note that what I am calling the "interactor question" does *not involve attributing adaptations or benefits to the interactors.* Interaction at a particular level involves only the presence of a trait at that level with a special relation to genic or genotypic expected success that is not decomposable into fitness components at another level.[8] A claim about interaction indicates only that there is an evolutionarily significant interaction occurring at the level in question; it says nothing about the existence of adaptations at that level. As we will see, the most common error made in interpreting many of the genetical models is that the presence of an interactor at a level is taken to imply that the interactor is also a manifestor of an adaptation at that level.

2.2. Replicators

Starting from Dawkins's view, Hull refined and restricted the meaning of "replicator," which he defined as "an entity that passes on its structure directly in replication" (1980, p. 318). I will use the terms *replicator* and *interactor* in Hull's sense throughout the rest of this chapter.

Hull's definition of replicator corresponds more closely than Dawkins's to a long-standing debate in genetics about how large or small a fragment of a genome ought to count as the replicating unit – something that is copied, and that can be treated separately (see especially Lewontin 1970).

[6] See Lloyd 1988, Ch. 4.

[7] For example, Arnold and Fristrup 1982; Brandon 1982; Colwell 1981; Damuth and Heisler 1988; Eldredge 1985; Griesemer and Wade 1988; Heisler and Damuth 1987; Lewontin 1970; Lloyd 1986c, 1988, 1989; Sober 1981, 1984; Sober and Lewontin 1982; Vrba and Gould 1986; Wade 1985; Wilson 1980; Wimsatt 1980, 1981.

[8] See the models cited in n. 5 for various technical approaches to expressing this special relation between fitness and trait.

This debate revolves critically around the issue of linkage disequilibrium and led Lewontin, most prominently, to advocate the usage of parameters referring to the entire genome rather than to allele and genotypic frequencies in genetical models (Lewontin 1970, 1974; Franklin and Lewontin 1970; Slatkin 1972; see discussion in Wimsatt 1980; Brandon 1982).

The basic point is that with much linkage disequilibrium, individual genes cannot be considered as replicators because they do not behave as separable units during reproduction. Although this debate remains pertinent to the choice of state space for genetical models, it has been eclipsed by concerns about interactors in evolutionary genetics.

2.3. *Beneficiary*

Who benefits from a process of evolution by selection? There are two predominant interpretations of this question: Who benefits ultimately, in the long term, from the evolution by selection process? And who gets the benefit of possessing adaptations as a result of a selection process?

Take the first of these, the issue of the ultimate beneficiary. There are two obvious answers to this question – two different ways of characterizing the long-term *survivors and beneficiaries* of the evolutionary process. One might say that the species or lineages (Hull's "evolvers") are the ultimate beneficiaries of the evolutionary process. Alternatively, one might say that the lineages characterized on the genic level, that is, the surviving alleles, are the relevant long-term beneficiaries. I have not located any authors holding the first view, but, for Dawkins, the latter interpretation is *the primary fact* about evolution. To arrive at this conclusion, Dawkins adds the requirement of *agency* (cf. Hampe and Morgan 1988). For Dawkins, a *beneficiary*, by definition, does not simply passively accrue credit in the long term; it must function as the initiator or causal source of a biochemical causal pathway. Under this definition, the *replicator* is causally responsible for all of the various effects that arise further down the biochemical pathway, irrespective of which entities might reap the long-term rewards.[9]

A second and quite distinct version of the "benefit" question involves the notion of adaptation. The evolution by selection process may be said to "benefit" a particular level of entity under selection, though producing

[9] The sense of agency assumed by Dawkins is worth investigating in detail. I will not, however, address this issue directly here. See Griesemer and Wade 1988. Related issues are discussed in Section 3.2.

adaptations at that level (Williams 1966, Maynard Smith 1976, Eldredge 1985, Vrba 1984). On this approach, the level of entity actively selected (the interactor) *benefits* from evolution by selection at that level through its acquisition of adaptations.

I think it is crucial to distinguish the question concerning the level at which adaptations evolve from the question about the identity of the ultimate beneficiaries of that selection process. One can think – and Dawkins does – that organisms have adaptations without thinking that organisms are the "ultimate beneficiaries" of the selection process.[10] I will therefore treat this sense of "beneficiary" that concerns adaptations as a separate issue, discussed in the next section, under the topic of the manifestor of adaptations.

2.4. Manifestor of Adaptations

At what level do adaptations occur? Or, as Sober puts this question, "When a population evolves by natural selection, what, if anything, is the entity that does the adapting?" (1984, p. 204).

As mentioned previously, the presence of adaptations at a given level of entity is sometimes taken to be a requirement for something to be a unit of selection.[11] Wright, in an absolutely crucial observation, distinguished group selection for "group advantage" from group selection per se (1980); in my terms, he claimed that the combination of the interactor question with the question of what entity had adaptations had created a problem in the group selection debates. Following Wright, I submit that the identification of a unit of selection with the manifestor of an adaptation at that level has caused a great deal of confusion in the units of selection debates in general.

Some, if not most, of this confusion is a result of a very important but neglected duality in the meaning of "adaptation" (in spite of useful discussions in Brandon 1978b; Burian 1983; Krimbas 1984, Sober 1984). Sometimes "adaptation" is taken to signify *any trait at all* that is a direct result of a selection process at that level. In this view, any trait that arises directly

[10] Brandon (1985) argues that such a view, which separates the level of adaptation from that of beneficiary, cannot be explanatory. Although I sympathize with Brandon's conclusions, they follow only under his set of definitions, which Dawkins and other genic selectionists would certainly reject.

[11] For explicit assumptions that being a unit of selection involves having an adaptation at that level, see Brandon 1982, 1985; Burian 1983; Mitchell 1987; Maynard Smith 1976; Vrba 1984.

from a selection process is claimed to be, by *definition*, an adaptation (e.g., Sober 1984; Brandon 1985, 1990; Arnold and Fristrup 1982).[12] Sometimes, by contrast, the term "adaptation" is reserved for traits that are "good for" their owners, that is, those that provide a "better fit" with the environment, and that intuitively satisfy some notion of "good engineering."[13] These two meanings, which I call the *selection-product* and *engineering* definitions, respectively, are distinct, and in some cases, incompatible.

Williams, in his extremely influential book *Adaptation and Natural Selection*, advocated an engineering definition of adaptation (1966). He believed that it was possible to have evolutionary change result from direct selection favoring a trait without having to consider that changed trait as an *adaptation*. Consider, for example, his discussion of Waddington's (1956) genetic assimilation experiments. Williams interprets the results of Waddington's experiments in which latent genetic variability was made to express itself phenotypically because of an environmental pressure (1966, pp. 70–81; see the lucid discussion in Sober 1984, pp. 199–201). Williams considers the question of whether the bithorax condition (resulting from direct artificial selection on that trait) should be seen as an adaptive trait, and his answer is that it should not. Williams instead sees the bithorax condition as "a disruption . . . of development," a failure of the organism to respond (1966, pp. 75–78). Hence, Williams draws a wedge between the notion of a trait that is a direct *product* of a selection process and a trait that fits his stronger *engineering* definition of an adaptation (see Gould and Lewontin 1979; Sober 1984, p. 201; cf. Dobzhansky 1956).[14]

This essential distinction between the *selection-product* and *engineering* views of adaptation is far from widely recognized. My claim here is that greater awareness of this distinction and its consequences will contribute to the understanding of several very heated debates in evolutionary theory.

[12] Oddly, Williams writes, "natural selection would produce or maintain *adaptation* as a matter of definition" (1966, p. 25; cf. Mayr 1976). However, Williams is committed to an engineering definition of adaptation (personal communication 1989).

[13] For example, Williams 1966; Bock 1980; Dunbar 1982; Ghiselin 1974; Gould and Lewontin 1979; Hull 1980; Lewontin 1978; Mayr 1978.

[14] Note that Williams says that "natural selection would produce or maintain adaptation as a matter of definition" (1966, p. 25; cf. Mayr 1976). This comment conflicts with the conclusions Williams draws in this discussion of Waddington; however, Williams later retracts this bithorax analysis (1985).

For example, the engineering notion of adaptation is at work in the long dispute over the relationship between natural and sexual selection. Many evolutionists, starting with Darwin, rejected the idea that the products of a sexual selection process should be considered *adaptations*. In fact, analysis of the process of sexual selection is sometimes motivated by the drive to find an explanation for the presence of "maladaptive" traits; hence, the distinction between the selection-product and engineering notion of adaptation plays an important role. Kirkpatrick (1987), for instance, uses a notion of adaptedness based on mean survival values in his argument that sexual selection does not always produce adaptations.

Consider for a moment the two schools of sexual selection theory. The "good genes" school claims that mate choice evolves under selection for females to mate with ecologically adaptive genotypes. The assumption here is that even though it appears that the females are basing their mate choice on a nonadaptive character, the character is actually an indication of the male's adaptedness (see, e.g., Vehrencamp and Bradbury 1984, Hamilton et al. 1990). The "nonadaptive" school claims that "preferences frequently cause male traits to evolve in ways that are *not adaptive* with respect to their ecological environment" (Kirkpatrick 1987, p. 44; emphasis added). In other words, the kinds of males preferred by females do not correspond with the kinds of males favored by natural selection. The result is a compromise between natural and sexual selection, the final state being one "that is maladaptive with respect to what natural selection acting alone would produce" (Kirkpatrick 1987, p. 45). Sir Ronald Fisher developed mathematical models showing how preferences for maladaptive males could evolve (1958 [1930]; see discussion in Lande 1980; Spencer and Masters 1992; Cronin 1991).

But an alternate concept of adaptation is available: the sexually selected traits that are advantageous to mating can still be seen as adaptations once the meaning of "adaptation" is adjusted. In this school of thought, the notion of "adaptation" should be broadened to include traits that contribute exclusively to reproductive success, even though the more traditional definition is in terms of engineering for survival (e.g., Cronin 1991, versus Bock 1980; see Kirkpatrick 1987).

As these perennial debates about the relation between sexual selection and natural selection show, when asking whether a given level of entity possesses adaptations, it is necessary to state not only the level of selection in question but also which notion of adaptation – either *selection product* or *engineering* – is being used. This distinction between the two meanings of adaptation also turns out to be pivotal in the debates about the efficacy of higher levels of selection, as we will see in Sections 3.1 and 3.3.

2.5. *Summary*

In this section, I have described four distinct questions that appear under the rubric of "the units of selection" problem: What is the interactor? What is the replicator? What is the beneficiary? and What entity manifests any adaptations resulting from evolution by selection? I also have discussed the existence of a very serious ambiguity in the meaning of "adaptation." I have no intention of defending one meaning or the other, but I will show that which meaning is in play has had deep consequences for both the group selection debates and the species selection debates. In Section 3, I will use my taxonomy of questions to sort out the most influential positions in three debates: group selection (Section 3.1), genic selection (Section 3.2), and species selection (Section 3.3).

3. AN ANATOMY OF THE DEBATES

3.1. *Group Selection*

Williams's famous near-deathblow to group panselectionism was, oddly enough, about benefit. He was interested in cases in which there was selection among groups and the group as a whole benefited from organism-level traits (including behaviors) that seemed disadvantageous to the organism.[15] Williams argued that the presence of a benefit to the group was *not sufficient* to establish the presence of group selection. He did this by showing that a group benefit was not necessarily a group adaptation.[16] His assumption was that a genuine group selection process results in the evolution of a group-level trait – a real adaptation – that serves a design purpose for the group. The mere existence, however, of traits that benefit the group is not enough to show that they are adaptations; in order to be an adaptation, under Williams's view, the trait must be an *engineering* adaptation that evolved by natural selection. Williams argued that group benefits do not, in general, exist *because* they benefit the group; that is,

[15] Similarly for Maynard Smith (1964).

[16] Hence, Williams is here using the term *benefit* to signify the manifestation of an adaptation at the group level.

they do not have the appropriate causal history (see Brandon 1981a, 1985, p. 81; Sober 1984, p. 262 ff.).

Implicit in Williams's discussion is the assumption that being a unit of selection at the group level requires two things: (1) having the group as an interactor, and (2) having a group-level engineering-type adaptation. That is, Williams combines two different questions, the interactor question and the manifestor-of-adaptation question, and calls this combined set *the* unit of selection question. These requirements for "group selection" make perfect sense given that Williams's prime target was Wynne-Edwards, who promoted a view of group selection that incorporated this same two-pronged definition of a unit of selection.

This combined requirement of "strong" (engineering) group-level adaptations in addition to the existence of an interactor at the group level is a very popular version of the necessary conditions for being a unit of selection within the group selection debates. David Hull claims that the group selection issue hinges on "whether entities more inclusive than organisms exhibit adaptations" (1980, p. 325). John Cassidy states that the unit of selection is determined by "Who or what is best understood as the possessor and beneficiary of the trait" (1978, p. 582). Similarly, Eldredge requires adaptations for an entity to count as a unit of selection, as does Vrba (Eldredge 1985, p. 108; Vrba 1983, 1984).

Maynard Smith (1976) also ties the engineering notion of adaptation into the version of the units of selection question he would like to consider. In an argument separating group and kin selection, Maynard Smith concludes that group selection is favored by small group size, low migration rates, and rapid extinction of groups infected with a selfish allele and that "the ultimate test of the group selection hypothesis will be whether populations having these characteristics tend to show 'self-sacrificing' or 'prudent' behavior more commonly than those which do not" (1976, p. 282). This means that the presence of group selection or the effectiveness of group selection is to be measured by the existence of nonadaptive behavior on the part of individual organisms along with the presence of a corresponding group-level adaptation. Therefore, Maynard Smith does require a group-level adaptation for groups to count as units of selection. As with Williams, it is significant that he assumes the *engineering* notion of adaptation rather than the weaker *selection-product* notion. As Maynard Smith puts it, "an explanation in terms of group advantage should always be explicit, and always calls for some justification in terms of the frequency of group extinction" (1976, p. 278; cf. Wade 1978; Wright 1980).

In contrast to the preceding authors, Sewall Wright separated the interactor and manifestor-of-adaptation questions in his group selection models (cf. Lewontin 1978; Gould and Lewontin 1979). Wright distinguishes between what he calls "intergroup selection," that is, interdemic selection in his shifting balance process, and "group selection for group advantage" (1980, p. 840; cf. Wright 1929, 1931).[17] He cites Haldane (1932) as the originator of the term "altruist" to denote a phenotype "that contributes to group advantage at the expense of disadvantage to itself" (1980, p. 840). Wright connects this debate to Wynne-Edwards, whom he characterizes as asserting the evolutionary importance of "group selection for group advantage." He argues that Hamilton's kin selection model is "very different" from "group selection for the uniform advantage of a group" (1980, p. 841; see Arnold and Fristrup 1982; Damuth and Heisler 1987, 1988).

Wright takes Maynard Smith, Williams, and Dawkins to task for mistakenly thinking that because they have successfully criticized group selection for group advantage, they can conclude that "natural selection is practically wholly genic." Wright argues, "none of them discussed group selection for organismic advantage to individuals, the dynamic factor in the shifting balance process, although this process, based on irreversible local peak-shifts is not fragile at all, in contrast with the fairly obvious fragility of group selection for group advantage, which they considered worthy of extensive discussion before rejection" (1980, p. 841).

This is a fair criticism of Maynard Smith, Williams, and Dawkins. My diagnosis of this problem is that these authors failed to distinguish between two questions: the interactor question and the manifestor-of-adaptation question. Wright's interdemic group selection model involves groups only as interactors, not as manifestors of group-level adaptations. Furthermore, he is interested only in the effect the groups have on organismic adaptedness and expected reproductive success. More recently, modelers following Sewall Wright's interest in structured populations have created a new set of genetical models that are also called "group selection" models and in which the questions of group adaptations and group benefit play little or no role.[18]

[17] It is worth remembering at this point that under Wright's view, interdemic group selection provided the means for attaining greater organismic adaptation; groups that are favored are "those local populations that happen to have acquired superior coadaptive systems of genes" (1980, p. 841).

[18] For example, Damuth and Heisler 1987; Heisler and Damuth 1988; Slatkin and Wade 1978; Uyenoyama 1979; Uyenoyama and Feldman 1980; Wade 1978, 1985; Wilson 1983.

For a period spanning two decades, however, Maynard Smith, Williams, and Dawkins did not acknowledge that the position they attacked, namely, that of Wynne-Edwards, is significantly different from other available approaches to group selection, such as Wright's, Wade's, Wilson's, Uyenoyama's, or Lewontin's. Ultimately, however, both G. C. Williams and Maynard Smith recognized the significance of the distinction between the interactor question and the manifestor-of-adaptation question. In 1985 Williams wrote, "If some populations of a species are doing better than others at persistence and reproduction, and if such differences are caused in part by genetic differences, this selection at the population level must play a role in the evolution of a species," while concluding that group selection "is unimportant for the origin and maintenance of adaptation" (Williams 1985, pp. 7–8).

And in 1987, Maynard Smith made an extraordinary concession:

> There has been some semantic confusion about the phrase "group selection," for which I may be partly responsible. For me, the debate about levels of selection was initiated by Wynne-Edwards' book. He argued that there are group-level adaptations . . . which inform individuals of the size of the population so that they can adjust their breeding for the good of the population. He saw clearly that such adaptations could evolve only if populations were units of evolution. . . . Perhaps unfortunately, he referred to the process as "group selection." As a consequence, for me and for many others who engaged in this debate, the phrase came to imply that groups were sufficiently isolated from one another reproductively to act as units of evolution, and not merely that selection acted on groups.
>
> The importance of this debate lay in the fact that group-adaptationist thinking was at that time widespread among biologists. It was therefore important to establish that there is no reason to expect groups to evolve traits ensuring their own survival unless they are sufficiently isolated for like to beget like. . . . When Wilson (1975) introduced his trait-group model, I was for a long time bewildered by his wish to treat it as a case of group selection, and doubly so by the fact that his original model . . . had interesting results only when the members of the groups were genetically related, a process I had been calling kin selection for ten years. I think that these semantic difficulties are now largely over. (1987, p. 123)

Dawkins also seems to have rediscovered the evolutionary efficacy of higher-level selection processes in an article on artificial life. In this article, he is primarily concerned with modeling the course of selection processes, and he offers a species-level selection interpretation for an aggregate species-level trait (Dawkins, 1989a). Still, he seems not to have recognized the connection between this evolutionary dynamic and the controversies surrounding group selection because in his second edition of *The Selfish Gene* (Dawkins, 1989b) he had yet to accept the distinction made so clearly by Wright in 1980. This was in spite of the fact that, by 1987, the importance of distinguishing between evolution by selection processes and any strong adaptations produced by those processes had been acknowledged by the workers Dawkins claimed to be following most closely, Williams and Maynard Smith.

3.2. Genic Selection

One may understandably think that Dawkins is interested in the replicator question because he claims that the unit of selection ought to be the replicator. This would be a mistake. Dawkins is interested primarily in a specific ontological issue about benefit. He is asking a special version of the beneficiary question, and his answer to that question dictates his answers to the other three questions under consideration in this chapter.

Briefly, Dawkins argues that because replicators are the only entities that "survive" the evolutionary process, they must be the beneficiaries. What happens in the process of evolution by natural selection happens *for their sake*, for their benefit. Hence, interactors interact for the replicators' benefit, and adaptations belong to the replicators. Replicators are the only entities with real agency as initiators of biochemical causal chains; hence, they accrue the credit and are the real units of selection.

Dawkins's version of the units of selection question amounts to a combination of the *beneficiary* question plus the *manifestor-of-adaptation* question. He does not somehow mistakenly think that he is answering the more predominant interactor question; rather, he argues that people who focus on interactors are laboring under a misunderstanding of evolutionary theory. One reason he thinks this, I submit, is that he takes as his opponents those who hold a combination of the interactor plus manifestor-of-adaptations definition of a unit of selection (e.g., Wynne-Edwards). Unfortunately, Dawkins thereby ignores those who are pursuing the interactor question alone; these researchers are not vulnerable to the criticisms he poses against the combined interactor-adaptation view.

I will discuss two aspects of Dawkins's own version of the units of selection issue. The first is his own preferred interpretation of the "real" units of selection problem. Here, I will attempt to clarify the key issues of interest to Dawkins and to relate these to the issues of interest to others. The second significant aspect of Dawkins's treatment of the units question is his characterization of the alternative views. The radical position Dawkins takes on units of selection makes more sense once his characterization of his opponents' questions becomes clear. In sum, he attributes to his opponents – sometimes incorrectly – a rich definition of a unit of selection involving not just the interactor question but also the beneficiary and manifestor-of-adaptation questions.

3.2.1. DAWKINS'S PREFERRED QUESTION: BENEFICIARY Dawkins believes that interactors, which he calls "vehicles," are not relevant to the units of selection problem. The *real* unit of selection, he argues, should be replicators, "the units that actually survive or fail to survive" (1982b, pp. 113–116). Organisms or groups as "vehicles" may be seen as the unit of function in the selection process, but they should not, he argues, be seen as the units of selection because the characteristics they acquire are not passed on (1982b, p. 99). Here he is following Williams's line. Genotypes have limited lives and fail to reproduce themselves because they are destroyed in every generation by meiosis and recombination in sexually reproducing species; they are only temporary (Williams 1966, p. 109). Hence, genes are the only units that survive in the selection process. The gene (replicator) is the real unit because it is an "indivisible fragment," it is "potentially immortal" (Williams 1966, pp. 23–24; Dawkins 1982b, p. 97).

The issue, for Dawkins, is "Whether, when we talk about a unit of selection, we ought to mean a vehicle at all, or a replicator" (1982b, p. 82). He clearly distinguishes the dispute he would like to generate from the group-versus-organismic selection controversy, which he characterizes as a disagreement "about the rival claims of two suggested kinds of vehicles" (1982b, p. 82). In his view, replicator selection should be seen as an alternative framework for both organismic and group selection models.

There are two mistakes that Dawkins is not making. First, he does not deny that interactors are involved in the evolutionary process. He emphasizes that it is not necessary, under his view, to believe that replicators are directly "visible" to selection forces (1982b, p. 176). Once again, his "vehicles" are conceived as the units of function in the selection process (like interactors) but not as the units of selection.[19] Dawkins has recognized from the beginning that his question is completely distinct from the interactor question. He remarks, in fact, that the debate about group versus

[19] Hull also has argued that his own "interactors" and Dawkins's "vehicles" are not the same things (1980).

individual selection is "a factual dispute about the level at which selection is most effective in nature," whereas his own point is "about what we ought to *mean* when we talk about a unit of selection" (1982a, p. 46). He also realizes that genes or other replicators do not "literally face the cutting edge of natural selection. It is their phenotypic effects that are the proximal subjects of selection" (1982a, p. 47). He suggests changing his own terminology to "replicator survival" to avoid this confusion but does not seem to have followed up.

Second, Dawkins does not specify how large a chunk of the genome he will allow as a replicator; there is no commitment to the notion that single genes are the only possible replicators. He argues that if Lewontin, Franklin, Slatkin, and others are right, his view will not be affected (see Section 2.2). If linkage disequilibrium is very strong, then the "effective replicator will be a very large chunk of DNA" (Dawkins 1982b, p. 89). We can conclude from this that Dawkins is not interested in the replicator question at all; his claim here is that his framework can accommodate any of its possible answers.

On what basis, then, does Dawkins reject the questions about interactors? I think the answer lies in the particular question in which he is interested, namely, What is "the nature of the entity *for whose benefit* adaptations may be said to exist" (1982b, p. 81, my emphasis; cf. pp. 4, 5, 52, 84, 91, 113, 114)?[20]

On the face of it, it is certainly conceivable that one might identify the beneficiary of the adaptations as – in some cases, anyway – the individual organism or group that exhibits the phenotypic trait taken to be the adaptation. In fact, Williams seems to have done just that in his discussion of group selection.[21] But Dawkins rejects this move, introducing an *additional* qualification to be fulfilled by a unit of selection; it must be "the unit that actually survives or fails to survive" (1982a, p. 60). Because organisms, groups, and even genomes are destroyed during selection and

[20] Compare an alternative formulation of Dawkins's central question: "When we say that an adaptation is 'for the good or something,' what is that something? . . . I am suggesting that the appropriate 'something,' the 'unit of selection' in that sense, is the active germ-line replicator" (1982a, p. 47). This particular formulation, I think, asks two questions, one about who the beneficiary of the selection process is and one about who possesses adaptations. Griesemer and Wimsatt's studies (1989) on Weismannism are of great help here.

[21] Note that Williams, even though he "keeps his books" in terms of genes, argued against the notion that particular group traits were group adaptations *because* these group traits are not properly understood as *benefiting the group* in the proper historical selection scenario (Williams 1966).

reproduction, the answer to the survival question must be the replicator. (Strictly speaking, this is false; it is *copies* of the replicators that survive. He therefore must mean replicators in some sense of information and not as biological entities [see Hampe and Morgan 1988; cf. Griesemer 2005]).

But there is still a problem. Although Dawkins concludes, "there should be no controversy over replicators versus vehicles. Replicator survival and vehicle selection are two aspects of the same process" (1982a, p. 60), he does not just leave the vehicle selection debate alone. Instead, he argues that we do not need the concept of discrete vehicles at all. I have shown elsewhere that if vehicles are understood strictly as interactors, Dawkins (and everyone else) *cannot* do without them (Lloyd 1988, Ch. 7).

In Dawkins's analysis, the fact that replicators are the only "survivors" of the evolution-by-selection process automatically answers the question of who owns the adaptations. He claims that adaptations *must* be seen as being designed for the good of the active germ-line replicator for the simple reason that replicators are the only entities around long enough to enjoy them over the course of natural selection. He acknowledges that the phenotype is "the all important instrument of replicator preservation," and that genes' phenotypic effects are organized into organisms (that thereby may benefit from them during their own lifetimes) (1982b, p. 114). But because only the active germ-line replicators survive, they are the true *locus of adaptations* (1982b, p. 113). The other things that *benefit* over the short term (e.g., organisms with adaptive traits) are merely the tools of the real survivors, the real owners. Hence, Dawkins rejects the vehicle approach partly because he identifies it with the manifestor-of-adaptation question, which he has answered by definition, in terms of long-term beneficiary.[22]

3.2.2. DAWKINS'S CHARACTERIZATION OF OTHER APPROACHES As discussed earlier, Dawkins is aware that the vehicle concept is "fundamental to the predominant orthodox approach to natural selection" (1982b, p. 116). He rejects this approach in *The Extended Phenotype*, claiming, "the main purpose of this book is to draw attention to the weaknesses of the whole vehicle concept" (1982b, p. 115). I will argue in the following paragraphs

[22] Mitchell arrives at a similar conclusion through different arguments. She highlights the impact of the notion of adaptation that seems to depend on the roles that Dawkins assigns to replicators (1987, pp. 359–362).

that his "vehicle approach" is *not* equivalent to what I have called the "interactor question" but encompasses a much more restricted approach.

In particular, when Dawkins argues against "the vehicle concept," he is arguing against the desirability of seeing the individual organism as the one and only possible vehicle. His target is explicitly those who hold what he calls the "Central Theorem," which says that *individual organisms should be seen as maximizing their own inclusive fitness* (1982b, pp. 5, 55). Dawkins's arguments are indeed damaging to the Central Theorem, but they are ineffective against other approaches that define units of selection more generally, that is, as interactors.

One way to interpret the Central Theorem is that it implies that the individual organism is always the beneficiary of any selection process; Dawkins seems to mean by "beneficiary" *both* the manifestor of an adaptation and that which survives to reap the rewards of the evolutionary process. He argues, rightly and persuasively, I think, that it does not make sense *always* to consider the individual organism to be the beneficiary of a selection process.

Note, however, that Dawkins is not arguing against the importance of the interactor question in general but rather against a particular definition of a unit of selection. The view he is criticizing assumes that the individual organism is the interactor, and the beneficiary, *and* the manifestor-of-adaptations. Consider his main argument against the utility of considering vehicles; the primary reason to abandon thinking about vehicle selection is that it confuses people (1982b, p. 189). But look at his examples; their point is that it is inappropriate always to ask how an organism's behavior benefits that organism's inclusive fitness. We should instead ask, says Dawkins, "whose inclusive fitness the behavior is benefiting" (1982b, p. 80). He states that his purpose in the book is to show that "theoretical dangers attend the assumption that adaptations are for the good of . . . the individual organism" (1982b, p. 91).

So, Dawkins is quite clear about what he means by the "vehicle selection approach"; he advances powerful arguments against the assumption that the individual is always the interactor cum beneficiary cum manifestor-of-adaptations. This approach is clearly *not* equivalent to the approach to units of selection I have characterized as the interactor question. Unfortunately, Dawkins extends his conclusions to these other approaches, which he has, in fact, not addressed. Dawkins's lack of consideration of the interactor definition of a unit of selection leads to two grave problems with his views.

One problem is that he has a tendency to interpret all group selectionist claims as being about beneficiaries and manifestors-of-adaptations as well as interactors; this is a serious misreading of authors who are pursuing the interactor question alone. Consider, for example, Dawkins's argument that groups should not be considered units of selection:

> To the extent that active germ-line replicators benefit from the survival of the group of individuals in which they sit, over and above the [effects of individual traits and altruism], we may expect to see adaptations for the preservation of the group. But all these adaptations will exist, fundamentally, through differential replicator survival. The basic beneficiary of any adaptation is the active germ-line replicator. (1982b, p. 85)

Notice that Dawkins begins by admitting that groups can function as interactors and even that group selection may effectively produce group-level adaptations. The argument that groups should not be considered real units of selection amounts to the claim that the groups are not the ultimate beneficiaries. To counteract the intuition that the groups do, of course, benefit, in some sense, from the adaptations, Dawkins uses the terms "fundamentally" and "basic," thus signifying what he considers the most important level. Even if a group-level trait is affecting a change in gene frequencies, "it is still genes that are regarded as the replicators which actually survive (or fail to survive) as a consequence of the (vehicle) selection process" (1982b, p. 115). Dawkins argues, "a population . . . is not stable and unitary enough to be 'selected' in preference to another population" (1982b, p. 100).

Saying all this does not, however, address the fact that other researchers investigating group selection are asking the interactor question and sometimes also the manifestors-of-adaptations question rather than Dawkins's special version of the (ultimate) beneficiary question. Dawkins gives no additional reason to reject these other questions as legitimate; he simply reasserts the superiority of his own preferred units-of-selection question.

This is fair enough, provided that Dawkins keeps the different questions clear. But he seems instead to misinterpret the claims of group selectionists. For instance, Dawkins believes that group selectionists hold an expanded version of the Central Theorem, that is, that group-inclusive fitness is "that property of a group which will appear to be maximized when what is really being maximized is gene survival" (1982b, p. 187). Although Wynne-Edwards might be characterized as holding this view, Wade and Wilson certainly cannot. I think that Dawkins rejects their projects because he does not distinguish them from Wynne-Edwards's program (1978, pp. 73–4; 1982b, p. 115).

The other, more serious problem is that Dawkins fails to address, in his own theory, the interactor question itself: Which entities can and should be delineated as having traits or properties by means of which they interact with the environment in ways that affect the process of evolution by natural selection? In his desire to eliminate empirically and theoretically unjustified claims about beneficiaries of the selection process, Dawkins omits consideration of relevant questions about phenotypes that are addressed by other theoreticians. His attempt to circumvent the problems of a very restricted approach to selection that focuses on the organismic phenotype leads him, unfortunately, to gloss over a gap in his own view – specifically, how to delineate the "extended phenotype."

In discussing the extended phenotype, Dawkins is interested exclusively in traits that "might conceivably influence, positively or negatively, the replication success of the gene or genes concerned" (1982b, p. 234). Other incidental traits are "of no interest to the student of natural selection"; therefore, they are not included in the extended phenotype (1982b, p. 207). This makes perfect sense; he focuses on traits on which selection operates. But does he offer a principle for identifying these traits? Sterelny and Kitcher (1988) managed to distill a method of determining significant phenotype out of Dawkins. It turns out, not surprisingly, that this method is a simple version of the same principle widely used to delineate *interactors* by those theorists working on the (pure) interactor question (Lloyd 1988, Ch. 7). Hence, Dawkins can accommodate a viable method for delineating the extended phenotype but only through landing himself squarely in the middle of the interactor debate.

In conclusion, Dawkins's objections are effective against a particular view of the units of selection in which the unit is an interactor plus the beneficiary plus the manifestor of an adaptation. These objections are ineffective, however, against the more sophisticated views about units of selection widely held among the geneticists developing hierarchical selection models based on the interactor question.

The tension between Dawkins's picture of his opponents and the actual range of views they hold has led to an unfortunate and severe case of arguing at cross-purposes. The genic selectionists never actually make contact with some of their supposed opponents, and the hierarchal modelers similarly fail to realize that the appropriately limited version of the genic selectionist claims is no threat to them. Dawkins's concern to establish the ontological priority of genes over the perceived theoretical hegemony of individual organisms falls on deaf ears among the geneticists, who do, after all, provide the inspiration for Dawkins's view in the first place.

3.3. Species Selection

Ambiguities about the definition of a unit of selection have also snarled the debate about selection processes at the species level. I argue in this section that a combination of the interactor question and the manifestor-of-adaptation question (in the engineering sense) led to the rejection of research aimed at considering the role of species as interactors, *simpliciter*, in evolution. Once it is understood that species-level interactors may or may not possess design-type adaptations, it becomes possible to distinguish two research questions: Do species function as interactors, playing an active and significant role in evolution by selection? And does the evolution of species-level interactors produce species-level engineering adaptations and, if so, how often?

For most of the history of the species selection debate, these questions have been lumped together; asking whether species could be units of selection meant asking whether they fulfilled *both* the interactor and manifestor-of-adaptation roles. For example, Vrba used Maynard Smith's treatment of the evolution of altruism as a touchstone in her definition of species selection (1984). Maynard Smith argued that kin selection can cause the spread of altruistic genes but that it should not be called group selection (1976).[23] Vrba agreed that the spread of altruism should not be considered a case of group selection because "there is no group adaptation involved; altruism is not emergent at the group level" (1984, p. 319; Maynard Smith gives different reasons for his rejection). This amounts to assuming that there must be group benefit in the sense of a design-type group-level adaptation. Vrba's view was that evolution by selection is not happening at a given level unless there is a benefit or adaptation at that level. That is, her definition of a unit of selection is a combination of an interactor and a manifestor-of-adaptations. She explicitly equates units of selection with the existence of an adaptation at that level (1983, p. 388); furthermore, it seems that she has adopted the stronger *engineering* definition of adaptation.

Eldredge also argues that species selection does not happen unless there are species-level adaptations (1985, pp. 196, 134). Eldredge rejects

[23] Again, this was because the groups were not interpreted as possessing design-type adaptations themselves.

certain cases as higher-level *selection processes* because "frequencies of the properties of lower-level individuals which are part of a higher-level individual simply do not make convincing higher-level adaptations" (1985, p. 133).

Vrba, Eldredge, and Gould all defined a unit of selection as requiring an emergent, adaptive property (Vrba 1983, 1984; Vrba and Eldredge 1984; Vrba and Gould 1986). In my analysis, this amounts to asking a combination of the interactor and manifestor-of-adaptations question.

But consider the lineage or species-wide trait of variability. Although this may come as a surprise to some, treating species as interactors has a long tradition (Dobzhansky 1956, Thoday 1953, Lewontin 1958). If species are conceived as interactors (and not necessarily manifestors-of-adaptations), then the notion of species selection is not vulnerable to Williams's original anti-group-selection objections.[24] The old idea was that lineages with certain properties of being able to respond to environmental stresses would be selected for, that the trait of variability itself would be selected for, and that it would spread in the population of populations. In other words, lineages were treated as interactors. The earlier researchers spoke loosely of adaptations, where adaptations were defined in the weak sense as equivalent simply to the outcome of selection processes (at any level). They were explicitly *not* concerned with the effect of species selection on organismic level traits but with the effect on species-level characters such as speciation rates, lineage-level survival, and extinction rates of species. I have argued, with Gould, that this sort of case represents a perfectly good form of species selection even though some balk at the thought that variability would then be considered, under a weak definition, a species-level adaptation (Lloyd and Gould 1993; Lloyd, 1988).

Vrba has also recognized the advantages of keeping the interactor question separate from a requirement for an engineering-type adaptation. In her more recent review article, she has dropped her former requirement that, in order for species to be units of selection, they must possess species-level adaptations (1989). Ultimately, her current definition of species selection is in conformity with a simple interactor interpretation of a unit of selection (cf. Damuth and Heisler 1988; Lloyd 1988).

It is easy to understand how the two-pronged definition of a unit of selection – as interactor and manifestor-of-adaptation – held sway for so

24 As Williams himself has acknowledged, in a discussion on species selection: "The answer to all these difficulties must be Lloyd's . . . idea that higher levels of selection depend, not on emergent characters, but on any and all emergent fitnesses" (1992, p. 27).

long in the species selection debates. After all, it dominated much of group selection debates until just recently. And I have argued that some of the confusion and conflict over higher-level units of selection arose because of an historical contingency –Wynne-Edwards's implicit definition of a unit of selection and the responses it provoked.

4. CONCLUSION

It makes no sense to treat different answers as competitors if they are answering distinct questions. I have offered a framework of four questions with which the debates appearing under the rubric of "units of selection" can be classified and clarified. I follow Dawkins, Hull, and Brandon in separating the classic question about the level of selection or interaction *(the interactor question)* from the issue of how large a chunk of the genome functions as a replicating unit *(the replicator question)*. I also separate the interactor question from the question of which entity should be seen as acquiring adaptations as a result of the selection process *(the manifestor-of-adaptations question)*. In addition, I insist that there is a crucial ambiguity in the meaning of *adaptation* that is routinely ignored in these debates: adaptation as a selection product and adaptation as an engineering design. Finally, I suggest distinguishing the issue of the entity that ultimately benefits from the selection process *(the beneficiary question)* from the other three questions.

I have used this set of distinctions to analyze leading points of view about the units of selection and to clarify precisely the question or combination of questions with which each of the protagonists is concerned. I conclude that there are many points in the debates in which misunderstandings may be avoided by a precise characterization of which of the units of selection questions is being addressed.

ACKNOWLEDGMENTS

I am grateful to the members of those audiences, and to Richard Dawkins, Steve Gould, Marjorie Grene, Jim Griesemer, Bill Hamilton, David Hull, Dick Lewontin, John Maynard Smith, Hamish Spencer, Mike Wade, and George Williams for helpful suggestions on previous drafts. A draft of this article was assigned to a Philosophy of Biology class at UC Davis (Winter 1991) by Michael Dietrich, and I thank him for his comments and suggestions. Some of this material was also presented informally in an interview in December 1989 with Werner Callebaut, parts of which were published in his *Taking the Naturalistic Turn: How Real Philosophy of Science Is Done (1993)*. An abbreviated presentation of the four basic questions (Section 2 of this chapter) is contained in my "Unit of Selection" essay for *Keywords in Evolutionary Biology* (1992), and I would like to thank Evelyn Fox Keller for insightful suggestions for improving the presentation of these ideas.

5

Species Selection on Variability

ELISABETH A. LLOYD AND STEPHEN JAY GOULD

1. SELECTION

We begin with some basic distinctions. Most important is the difference between a process and an outcome of that process; in this case, selection is the process, and evolutionary change is the outcome. The process of selection involves an interaction between an entity and its environment or, more specifically, between a particular trait of an entity and particular aspects of its environment. Part of the fitness calculated for an entity represents the particular value conferred by having a specific trait in a specific environment. In other words, the component of fitness associated with a trait is the representation, in an evolutionary model, of the process of selection.

The selection process itself must be contrasted with the result or outcome of that process. Sorting is simply the differential birth and death of individual entities in a population, while selection is a potential cause of that sorting (Vrba and Gould 1986, p. 217). Sorting at a particular level has a variety of potential causes; it can arise from selection acting at that level or occur as a consequence of chance or of selection processes at either a higher or a lower level.

2. SPECIES SELECTION

A number of attempts have been made to apply the basic notion of selection to the species level. Vrba, Eldredge, Gould, and others (Vrba and Gould 1986; Vrba 1984; Vrba 1989; Vrba and Eldredge 1984; Stanley

Lloyd, E. A., and Gould, S. J. (1993). Species Selection on Variability. *Proceedings of the National Academy of Sciences, 90*, 595–599.

1975; Stanley 1979; Damuth and Heisler 1988) in several articles have defended the idea that selection at a particular level requires characters to be heritable and emergent at that level and to interact with the environment to cause sorting. Species-level properties are divided into aggregate and emergent characters. Aggregate characters are based on the inherent properties of subparts and are simple statistics of these properties, while emergent characters arise from the organization among subparts (Vrba and Eldredge 1984, p. 146). Possible emergent characters among species include population size, distribution, and composition (Vrba and Eldredge 1984).

According to this argument, which we shall call the "emergent character" approach, one need only show that a trait in question is aggregate, not emergent, in order to demonstrate that species are not functioning as units of selection in a particular case (Vrba and Gould 1986; Vrba 1984; Vrba and Eldredge 1984).

We shall contrast the emergent character approach with an "emergent fitness" or "interactor approach," which emphasizes the nature of fitness values, rather than type of character. Under this approach (presented in more detail in Section 4), the relationship between character and environment becomes the key issue. This relationship is represented in the fitness parameters. Selection processes are delineated by distinctive attributes of the fitness parameters. Interactors, and hence selection processes themselves, are individuated by the contributions of their traits to fitness values in evolutionary models; the trait itself can be an emergent group property or a simple summation of organismic properties. This definition of an entity undergoing selection is much more inclusive than in the emergent character approach, because an entity might have either aggregate or emergent characters (or both); this distinction between emergent and aggregate characters does not usually appear in quantitative models of species selection (e.g., see Slatkin 1981). The emergent fitness approach requires only that a trait have a specified relation to fitness in order to support the claim that a selection process is occurring at that level (cf. refs. Damuth and Heisler 1988; Lewontin 1970; Arnold and Fristrup 1982; see Section 4).

3. ADAPTATIONS

An important component of the emergent character approach is its focus on adaptations. The existence of an emergent character at the species level is taken as equivalent to the existence of a potential species-level adaptation (Vrba 1983, p. 388; Eldredge 1985, p. 132).

The focus on characters rather than fitnesses is related to this requirement for adaptations. Emergent characters are always potential adaptations. Not all selection processes produce adaptations, however. The key issue, in delineating a selection process, is the relationship between a character and fitness. The emergent character approach is more restrictive than alternative schemas that delineate selection processes, for only some selection processes at the species level (perhaps very few) result in the evolution of emergent characters or adaptations. The emergent character approach picks out a subset of cases given by the interactor approach; it is, to be sure, an especially interesting subset – namely, species-level adaptations. Nevertheless, if one is interested in discussing evolution by species-level selection processes *per se*, then this approach is too restrictive.[1] Let us return to the emergent fitness approach to see how aggregate traits, which are not adaptations, could function in a species selection process.

4. EMERGENT FITNESS OR INTERACTOR APPROACH

As the guiding idea behind our hierarchical view of selection models, we hold that selection processes are best described in terms of interactors (see Hull 1980; Brandon 1982; Dawkins 1982). An interactor is an entity that has a trait; the interactor must interact with its environment through the trait, and the interactor's expected survival and reproductive success are determined, at least partly, by this interaction. In other words, the interactor's fitness covaries with the trait in question.

If we have an interactor, we should expect to find a correlation between the interactor's trait and the interactor's fitness.[2] For evolution by

[1] Dawkins's arguments against species selection are refreshingly clear about this distinction. Following his admission that species selection is, in principle, possible, he expresses skepticism that species selection may be important in explaining evolution. He then continues "This may just reflect my biased view of what is important . . . what I mainly want a theory of evolution to do is explain complex, well-designed mechanisms like hearts, hands, eyes, and echolocation" (1986, p. 265). We would reply that a theory of evolution has many more, and equally important, things to do. Dawkins may be revising his view, because he has now offered a species-level selection interpretation for an aggregate species-level trait (Dawkins 1989).

[2] Several methods are available for expressing such a correlation. For instance, under the covariance approach, the change in the mean value of a trait under selection can be expressed as a covariance between relative fitness and the character value (i.e., quantitative

selection to occur, both the trait and the correlation between trait and fitness must be at least partly heritable (in the sense of narrow heritability). Heritability depends, in turn, on the additive genetic variance of the trait by definition. It should be recalled that fitness itself can be analyzed into components, each correlated with a trait that affects fitness. In selection models, the interactions *per se* of trait and environment are not represented; rather, the evolutionary effect of this interaction is represented by the selection coefficient (fitness parameter). We can model the selective effects of interactions by partitioning the overall fitness into levels where the proper correlation between a trait and a component of fitness can be represented. In selection models, then, those interactions between trait and environment that yield evolutionary changes are represented by an additive component of variance in fitness correlated with variance in the trait in question.[3] In considering a hierarchy of selection models, we are simply generalizing this principle that relates the efficacy of natural selection to additivity. Note that this view can accommodate more than one level of selection operating simultaneously.

Clearly, a potential problem exists here. Suppose the correlation between trait and fitness at a higher level is a simple effect of the traits and fitnesses at a lower level. Because we do not want to count these lower-level interactions twice, we must avoid representing selection at the higher level in this case. This is done by describing interactors at the lower level first. If a higher-level interactor exists, the higher-level correlation of fitness and trait will appear as a residual fitness contribution at the lower level; we must then go to the higher level in order to represent the correlation between higher-level trait and higher-level fitness. Hence, species-level fitness is not defined as an average or sum of the organismic fitnesses within a species (see Damuth and Heisler 1988; Arnold and Fristrup 1982; Lloyd 1988).

More simply put – the operative notion in all examples of higher-level selection is some sort of interaction effect or context dependence (e.g., Wimsatt 1981; see Sober 1984, Ch. 7; here, "context" refers to the other

description of the trait). Actually, it is somewhat imprecise to refer to "an interactor's fitness"; this quantity should be understood as the probability associated with the interaction of the interactor and its environment.

[3] Damuth and Heisler (1988) claim "selection can be occurring at a level only if there is a relationship between some character of the units at that level and the fitnesses of those units." They assess the existence and intensity of this relationship through the regression of the fitness on the character value (see the selection gradient analysis, a phenotypic approach [Lande and Arnold 1983]).

entities in the same population and not to the environment). Intuitively, the properties of the other members of your group must make a difference, and there must be a correlation between group type and fitness.

One might object that the above approach to delineating interactors is overly reductionist. After all, it seems that this view simply embodies G. C. Williams's famous and misleading maxim: don't even consider any higher level of selection unless the lower-level model proves empirically inadequate (Williams 1966, p. 55). The problem with Williams's maxim is that different selection processes may sometimes yield identical gene frequency predictions. In these cases, additional information about population structure, group membership, and group-level fitnesses is needed in order to tell which model best fits the system at hand (see Lloyd 1988, pp. 86–96).

The approach that we support is intended not as a research strategy, as Williams's maxim has often been used, but as a method of evaluation or calculation. Our recommended method for assessing selective levels is used only after all the information has been obtained, including data about population structure, group composition, and group-level fitnesses. In the method we advocate, and as a crucial difference from Williams's approach, this information is not collected under a reductionist research program; for in such a program, the lowest-level selection model is accepted without even testing whether a higher-level model might be more adequate (see Dawkins 1982; Lloyd 1988; Wimsatt 1981; Hamilton 1967; Wimsatt 1980; Wilson 1983).

5. VARIABILITY

Dobzhansky, in his *Genetics and the Origin of Species* (1937), considered variability as a species-level character related to species survival (cf. refs. Darlington 1939; Simpson 1953; Thoday 1953; Nicholson 1960). Dobzhansky notes that a reservoir of genetic variation within species acts as a hedge against extinction (Dobzhansky 1937, p. 127). He sees a trade-off involved in variability; species that concentrate adaptations very narrowly are favored by natural selection at a given moment, but they sacrifice plasticity, "the flexibility that retention of a goodly amount of genetic variation affords against the (inevitable) change in position of the adaptive peak" (Eldredge 1985, p. 199; Slatkin 1981; cf. Lewontin 1970, p. 15; Slobodkin and Rapoport 1974).

Eldredge acknowledges that the reservoir of genetic variability discussed by Dobzhansky could play a role in species survival (Eldredge 1985, p. 182; see Fowler and MacMahon 1982). But although Lewontin (1958) talks of selection contributing to or maintaining variation, Eldredge claims that variation is not being selected because it conveys an advantage; rather, the maintenance of variation is just an effect of ordinary selection on organisms. Although Eldredge agrees that variability is correlated with long-term species survival, he rejects the possibility that this might count as species selection because variability is not an adaptation. We claim, in contrast, that variability is a perfectly good species-level trait that can be associated with genuine species-level fitness (see Lloyd 1988, pp. 110–112).

Many evolutionists shared Dobzhansky's interest in variability and its involvement in long-term species survival. During the early 1950s, evolutionists produced genetic models of the relation between variability and what were then called (improperly, by our definitions) species adaptations (Thoday 1953; Lewontin 1958). Why was this subject dropped? We suggest that the virtually universal failure to distinguish between adaptations and selection processes led to this abandonment. The attack on group adaptations, led by Williams (1966), was interpreted as an assault on the possibility of group-level selection processes in general, although such a conclusion does not follow. Hence, we suggest, species-level variability, like other group-level properties, was discredited through its association with unsupportable arguments for group-level adaptationism (Fowler and MacMahon 1982, p. 491). We would like to revive interest in the relation between genetic variability across groups and lineages, and long-term success of lineages, by placing this subject in the context of hierarchical selection theory.

Williams did a service to the community of evolutionary biologists by refining the definition of adaptation and by insisting on strict standards of proof for adaptedness. His analysis drove a wedge between fitness (expected reproductive success) and adaptedness, where only adaptedness signifies adaptation (Brandon 1981b; Gould and Vrba 1982). Unfortunately, biologists seem to have lost track of those selection processes that do not yield cumulative adaptations. (A stronger link may exist between adaptation and selection at the organism level than at higher levels. This might constitute a genuine and interesting difference between these levels.) If evolutionary theory is to yield accurate models of all selection processes, however, then the set of available models must be expanded.

Once the definition of species selection has been brought into alignment with other hierarchical selection models, species selection gains potential for a greatly expanded role in evolution.[4] For example, let us turn to the key item barred by the emergent character approach but included under the emergent fitness approach–variability. We next offer a reinterpretation of a classic case presented in support of species selection. The original interpretation, advanced by Gould, is vulnerable to a counter-interpretation at the organism level. When the case is recast in terms of variability, however, the organism-level reinterpretation loses its force.

6. NEOGASTROPODS

Cases that suggest no clear adaptive explanation for trends in organismal phenotypes are promising for species selectionists. Trends may occur because some species speciate more often than others. For example, many clades of marine invertebrates exhibit a trend toward increased frequency of stenotopic species (stenotopes are narrowly adapted to definite environmental factors; eurytopes can tolerate a broad range of environments; see Vrba 1980; Gould 1982; Hansen 1978; Hansen 1980; Jablonski 1986; Jablonski 1987). Sometimes this trend leads to elimination of eurytopes completely, as in the volutid neogastropods (Hansen 1978). Analysis of this clade shows a much greater rate of speciation among stenotopic species, which helps to produce the trend even though stenotopes also suffer a higher rate of extinction.

The higher rate of speciation in stenotopes has been interpreted as resulting from isolation of small populations; speciation may be enhanced because stenotopes generally brood their young, while eurytopes tend to have planktonic larvae – and species with planktonic larvae exhibit high levels of gene flow (Gould 1982, p. 96). Jablonski (1986) argued that different modes of larval development confer different population

4 Discussions of the evolution of sexual reproduction have sometimes involved species- and lineage-level selection processes. Stanley, for example, argued that a sexually reproducing ancestral species produces a much greater ecological diversity of descendant species than an otherwise identical asexual species; this greater diversity makes the sexual lineage less vulnerable to extinction than the asexual one (cf. Stanley 1975; Ghiselin 1974; Maynard Smith 1978; Michod and Levin 1988; Williams 1975). Vrba, who formerly emphasized the limited role of species selection in evolution, based on the requirement that it must involve species-level adaptations (Vrba 1984; Vrba 1989; Vrba 1983), now thinks that species selection will be "quite common" (Vrba, 1989, p. 162).

structures, which could be counted as species-level traits. (We part with Jablonski's analysis in that he supported the claim for species selection by characterizing larval ecology as an "emergent property," whereas we concentrate on tying the population structure to lineage-wide genetic variability.) The trend is interpreted as a case of species selection because stenotopic species take over the clade by differential speciation, which is not sensibly explained by organismic-level natural selection (Gould 1982, p. 97; see discussion in Arnold and Fristrup; Jablonski 1986).

But why call this species selection? The trend toward stenotopy in the lineage can be seen as an effect of an organismic-level property – namely, the tendency for larvae to be nonplanktotrophic. The key factor in the higher rate of speciation is isolation, and a primary cause of isolation is the feeding habits of individual offspring. Hence, a species-level pattern is produced but does not arise from a true species-level property. In fact, those who require a species-level adaptation would certainly reject non-planktotrophy, as it is a property of organisms. Therefore, the organismic selectionist might argue, there is no species selection operating here; this is just a case of organismic-level selection producing a higher-order effect (e.g., see Dawkins 1986, p. 266).

In response to this challenge, we would like to recast this case in terms of variability and its role in the evolution of lineages. Consider the case in which eurytopes have been eliminated from the lineage: A species with nonplanktotrophic young has a tendency to develop isolated populations – hence, a tendency to speciate. Speciation leads (in this case) to more genetic variability across the lineage (counting from some ancestral gene pool; see Thoday 1953). These species also have a relatively high probability of becoming extinct. [Jablonski (1987) provides additional support for the argument that variability provides a target for species selection. He documents the role of geographic range in species persistence and heritability.] But, for the overall lineage, enhanced speciation leads to success.[5] Note that the immediate cause of success is the tendency to speciate, but the long-range explanation of success could be related to variability.

[5] As a rough notion of success, we can borrow Thoday's definition of the fitness of a "unit of evolution," which is equal to the probability that it "will survive for a given long period of time, such as 10^8 years," – that is, it "will leave descendants after that lapse of time" (Thoday 1953, p. 98). Similarly, Arnold and Fristrup define success as a measure of the relative increase or decrease in descendants of a lineage (and fitness as the expectation of that success; Arnold and Fristrup 1982, p. 120; cf. Cooper 1984).

The species with planktotrophic young do not speciate much, as a result (partly at least) of extensive gene flow; hence, they present fewer alternative strategies for facing environmental challenges. These species are longer-lived relative to the stenotopes – but note that, in the case of the volutes, the eurytopes are all extinct (Stanley 1975).

Let us reconsider the trait, "tendency to speciate." Gould (1982) originally considered this feature both *(i)* an emergent property, and *(ii)* a property resulting from a larval strategy – that is, an individual-level trait. But in this case, stenotopes not only tend to speciate more; the extinction rate, even though higher than in eurytopes, is also lower than the speciation rate. Hence, stenotopes are the only long-term survivors in the lineage. That is, there is advantage (to the original gene pool) in speciating more. We propose that the advantage comes from success in some of the experiments that occur during speciation. If one counts the total genetic variability across the whole lineage, then stenotopes as a collective lineage maintain more variability than eurytopes. This variability is expressed in the speciation experiments, and the presence and maintenance of variability contribute to the long-term success of the stenotopic lineages.

This is not the only possible evolutionary scenario that could be given for this case, of course, but it is a possible and testable candidate. A lineage-level component of fitness based on variability can, in principle, be quantified and entered into the models previously discussed. By taking this alternative seriously, we avoid thinking that simply because a higher-level trait in a species is, in fact, caused by the behavior of individual organisms, species selection cannot be operating (Lloyd 1988, Ch. 7; Mayo and Gilinski 1987; Gilinksi 1986). Hence, in our view, a trait can ordinarily be considered as a genic- or organismic-level trait, but it can nevertheless participate causally in species selection. The existence of such cases will expand the amount of evolutionary change explicable by species selection.

7. MACROEVOLUTIONARY SIGNIFICANCE OF SPECIES SELECTION ON VARIABILITY

We have shown, by an argument based on a logical analysis of evolutionary theory, that variability and other aggregate traits can figure in species selection defined by emergent fitnesses under the interactor approach.

But although cogent logic may define the philosopher's task, it can only represent the starting point for a practicing biologist. The authors of this paper, as a collaboration of both professions, must therefore pose a further question: is species selection on variability important in evolution; does it display a high relative frequency among the causes of trends? Almost all major questions, and great debates, in natural history revolve around the issue of relative frequency: for example, selection and neutrality, adaptation and constraint.

The centrality of this point was recognized by Fisher (1958). He acknowledged that the logic of species selection was unassailable but denied this process any important role in evolution by arguing that it would always be overwhelmed by his (and Darwin's) favored mode of selection on organisms. The number of organism births overwhelms species births by so many orders of magnitude, Fisher argued, that nothing much can accumulate by selection among species relative to organisms.

This argument might work in Fisher's world of uniformity, isotropy, and universally effective (and therefore nonrestrictive) intraspecific variability. But the geological stage of macroevolution presents a situation most uncongenial to Fisher's assumptions – a world that suggests an important relative frequency for selection on species-level variability. Consider just two aspects of the fossil record.

(i) Punctuated equilibrium. Despite continuing arguments about interpretation (Eldredge and Gould 1972; Gould and Eldredge 1977; Turner 1986; Gould 1989), nearly two decades of study and debate have established a high relative frequency for the geometry and topology underlying this theory (Cheetham 1986; Stanley and Yang 1987) – geologically abrupt appearance and later stasis of most morphospecies in the fossil record. Fisher's argument fails because most species may be constrained to remain stable, thus rendering irrelevant the orders of magnitude advantage of individual births, and making rare events of speciation "the only game in town." Trends must therefore arise as differential success of species, and clades with greater interspecific variability as a result of more copious speciation, may gain a macroevolutionary edge. (Our explanation for volute evolution in the last section presents an argument in this mode.)

(ii) Mass extinction. If differential removal of species during rapid and worldwide episodes of mass extinction sets the basic pattern of life's diversity through time, then failure to maintain variation becomes an especially potent cause of species death, because survival through unanticipated environmental challenges of such magnitude must often depend on fortunate success of a few variants, whereas narrow adaptation and limited variability must often lead to elimination. Thus, in the actual and uncertain world that geology has set for the history of life, differential success of species must regulate many trends – and variability across species within clades must be a major component of success or failure. Species selection on variability is probably a major force of macroevolution.

6

An Open Letter to Elliott Sober and David Sloan Wilson, Regarding Their Book, *Unto Others*

The Evolution and Psychology of Unselfish Behavior

Dear Elliott and David,

I shall start with an overview of how I see things quite generally, and what my strategy has been in approaching the units of selection debates. Then I shall address individual points regarding your book.

1.

For the past ten years, I have been focusing my attention in the units of selection debates on a *diagnostic* question. The situation, as I was convinced by the time of the publication of my 1988 book (*The Structure and Confirmation of Evolutionary Theory*) by both your work, and the work of Michael Wade, Robert Colwell, Marcus Feldman, Montgomery Slatkin, and others, is that good models and good evidence exist to demonstrate the presence and efficacy of levels of selection above the individual organism or mating pair. It also seemed clear to me that some of these higher-level cases were not best understood as families (i.e., as subject to kin selection): in Wade's case because the cases weren't, and in Wilson and Colwell's case because I was persuaded by their arguments and evidence that kin selection is a special case of group selection.

Yet many evolutionists remained unconvinced by any of the evidence and models that I found compelling. This led me to a question: Why did John Maynard Smith, George C. Williams, Len Nunney, Robert Trivers, Richard Dawkins, and so on, continue to argue against the group selection models and the evidence? What standard of evidence or what requirement were they using, that these models and this evidence – which I found so compelling – failed to meet?

I proposed an answer to these questions in my 2001 paper, "Units and Levels of Selection." There, I argued that these opponents are defining group selection (although usually not explicitly) in such a way that these higher-level models and experiments do not qualify as such. Your reaction has been to insist that these are genuine cases of group selection, and I agree with you. But I am also interested in this other diagnostic question, of why the opponents think that the cases you presented are *not* instances of group selection. In my 2001 paper, I offered answers to this and other puzzles, such as: why they are so committed to the idea that group selection is rare or impossible; why they always insist that selection must be operating as kin selection; and why they insist that all effective cases of group selection are really cases of kin selection (a mantra at Oxford in the late 1980s).

My idea is very simple, and it takes seriously what these opponents to group selection have been saying all along. They really do have something else in mind when they talk about "group selection." I agree with you that they should not have something else in mind, and have argued this with each of them personally, as well as in print. But the first step in adjudicating this debate is getting clear about what we disagree about.

Thus I analyzed the units of selection question into combinations of four different questions, each of which is perfectly clear and conventional: the interactor, replicator, manifestor-of-adaptation, and beneficiary questions, respectively. In addition, I identified two different notions of adaptation that were prevalent in the literature and dubbed them "strong" or "engineering" adaptation and "weak" or "product-of-selection" adaptation. Many participants have talked about units of selection in one or more of these ways; the application of my analysis involves investigating which *combination* of requirements the participants of the debate have in mind when they say that such-and-so is or is not a case of group selection.

I want to emphasize that these four definitions of a unit of selection are diagnostic categories; making these distinctions does not involve endorsing any of them. In fact, I remain neutral on whether or not the strong or weak notion of adaptation should be used. I do argue, however, that only the interactor question is the pure, dynamical units of selection question about the level at which a selection process occurs (1988, Chs. 4–5).

Distinguishing between these four units of selection questions requires, in part, that one sort out an equivocation in the usage of the concept of adaptation. On the "engineering" (or "design") view of adaptation, an adaptation is a novel, evolved mechanism that "solves" a problem faced by the manifestor-of-adaptation in its environment. Note that a trait cannot be "designed" with respect to a selection process, in this sense, unless it is a novel trait, but that some novel traits are not "designed," in this sense. On the product-of-selection view, an adaptation is *any* aspect of the interactor involved in a natural selection process. Because not all natural selection processes result in "design" type adaptations, the product-of-selection view encompasses many more results of evolutionary processes (Spencer 2001). To give an example of a product-of-selection adaptation, consider Kettlewell's moth case. There, the population started with two distinct morphs. Selection on the two morphs resulted in a new, higher frequency of dark morphs in the population. Because this is the paradigmatic form of natural selection in the wild, it is important to notice that the result of this selection process was not a new trait, and therefore not a design-type adaptation at all, but merely a change in frequency of different phenotypes. Thus, we have a good case of selection which produces an evolutionary change, but not one that produces new machinery, that is, an engineering adaptation. Certainly there are mechanisms underlying the different phenotypes, and these were selected, but no new morph or machinery resulted from this selection process.

Note the crucial point that the logic of the two views of adaptation and their relation to selection processes is different. Only the product-of-selection view supports the biconditional: there is selection if-and-only-if there is adaptation. The engineering view only supports the claim: if there is an engineering adaptation, then we must infer that selection occurred.

One of the most important issues in the group selection debates (to your opponents) – but one that is hidden in plain sight because of the terminological ambiguity – is whether and how often groups evolve engineering adaptations. Look at Dawkins: his obsession is with the evolution of complex adaptations, ones accumulated over generations through complex collections of genes – adaptations that are like machines to solve specific engineering problems "posed" by the environments of the organism (or the gene). You both and I may look at this and think: yes, that happens, that is an important outcome of some evolutionary processes, but it is not the same as a selection process itself. After all, we know that selection sometimes does and sometimes does not lead to the accumulation of complex adaptations, things that act as functionally integrated components, things that could look like they were designed to serve a particular function.

Thus, we can see that when Dawkins opposes group selection because it doesn't lead to group "benefit" or group adaptation, *he doesn't mean the same thing by "adaptation" as you do.* You use the term to refer to any outcome of a selection process (what I call "weak" or "product-of-selection" adaptation). He uses the term to refer to complex evolved functionally integrated "designed" machines. So if you offer him models and evidence that show that group selection operates, and produces product-of-selection adaptation – which is precisely what you argue for in your book – then it makes sense that he then turns around and says, No, you haven't given me evidence that group selection produces (engineering) adaptation. And he is right, you have not.

What is required at this point is an argument that engineering adaptation is the *wrong standard to use* to establish whether group selection has or has not occurred (I have given such an argument in my 1988 book, Chs. 5–7). I was delighted to hear that when David Wilson went to Oxford, the folks there talked with him about the different standards of evidence for engineering and product-of-selection adaptation, and whether and how each model or case supports each of the different claims for (1) the existence of a selection process, and (2) the existence of an engineering-type adaptation. This is the very distinction that I have been trying to get them to consider for ten years. Whether they have accepted the argument that group selection involves only the dynamical process of selection, and does not require the existence of accumulated engineering adaptations, remains to be seen, I guess, although Maynard Smith did agree to this in private conversation.

So, it is with this background set of analyses and concerns that I came to your book. What I found was an excellent and compelling summary of the empirical evidence supporting the presence and efficacy of group selection processes in nature. This means that I found very good evidence for group adaptation in the product-of-selection sense, that is, evolutionary changes in group representation depending on phenotype.

What I did not see (and this is not a problem for me, because I do not think it is relevant to a discussion of group selection *per se*) is good evidence for the existence of engineering-type group adaptations. I do think that the altruism, social norms and punishments, and two-layer reinforcing selection scenarios you discuss are highly suggestive, and that they promise that it is reasonable to *expect* complex, engineering adaptations in social species. Nevertheless, for the hard-core, they are not in the bag. I have been talking with these opponents of "group selection" (really: of "engineering adaptation by group selection") for ten years about what kinds of evidence or models would be required to show engineering-group-adaptation, and you have not presented enough in the book to persuade even me, and I'm inclined to believe.

Given that you yourselves explicitly adopt the product-of-selection view of adaptation, there is no internal reason that you should even have addressed this engineering issue. Elliott, your comments to me in Oaxaca suggested that you might want simply to reject the coherence of the engineering definition itself. But that is a different job, and you would be up against many folks, from Darwin through Richard Lewontin, Stephen Jay Gould, and Ernst Mayr. The problem that I am trying to highlight is that, even if you do not adopt the engineering definition, *others are using it against you*. So I think this means you have to address the problem head-on, either by claiming that you are not defending any claims about strong (engineering) group adaptations, or by arguing versus the entire tradition in evolutionary biology that thinks that engineering adaptations are an important topic of investigation, or by coming up with the goods in the form of strong evidence for group-level engineering adaptations that are not correctly interpreted any other way.

Elliott, your remarks in our discussion in Oaxaca suggested to me that you were inclined toward the "not defending any strong group adaptations" strategy. In fact, given your explicit adoption of the product-of-selection version of adaptations (1984), this is the natural reading of your new book. I was led, though, by the use of terms like "functional integration" and "complexity" and "organism-like" to think that you sometimes wanted to address more than simple product-of-selection adaptation; these are the classic ways to refer to engineering adaptations. Given your opponents' obsessions with engineering adaptations, I am sure that they will have picked up on this use of language, and think, as I did, that you are sometimes making claims for more than product-of-selection adaptations.

2.

Now I shall address more specific aspects of your book. First, I claim in my review of the book (1999, p. 447) that "the most significant part of the book for practicing biologists" is that W. D. Hamilton shifted to a multilevel selection point of view, but that his followers failed to notice this. I emphasize this because the audience of the review includes the sizable portion of evolutionary geneticists and evolutionary ecologists who go around saying: "but Hamilton has given us the most powerful modeling tool of all time, inclusive fitness theory, which is our favorite way to explain everything without ever having to consider groups." Every major review of *Unto Others* that I have read said this very thing. Of course, Hamilton's own change of mind does not really matter, in the ideal world, but this is the actual world, where people who think that Hamilton is the most important geneticist of his generation might be influenced by his change of mind in a way that they will not be influenced by anything else.

I was especially struck by the fact that these reviewers repeated the standard mantra, above, including the citing of Hamilton, without apparently having even realized that they were disagreeing with Hamilton himself. I think they need to be forcefully reminded of his change of mind; they are members of a school that claims to represent him.

Now, here are my renditions of objections that you have raised with me about my review of your book, and some replies:

1. Your Objection: Groups and their traits must be defined independently of the outcome of multilevel selection, and this is done by defining groups in terms of fitness effects, and not in terms of group-level adaptations in the engineering sense.

Response: I hope it is clear now, that I am not in favor of defining groups in terms of engineering group adaptations. I have argued against it, and in favor of defining group traits in terms of fitness effects (1988/1994, Ch. 5).

2. Your Objection: With respect to the definition of adaptive groups, the distinction between "weak" and "strong" adaptation is a recent development. You suggest that perhaps I am historically incorrect when I say "G. C. Williams, Maynard Smith, and other opponents of group selection were arguing against the production of group adaptations in the strong, engineering sense." You note that *all* of the classic examples of group selection that were debated in the 1960s and 1970s, including sex ratio, virulence, altruism, and population regulation, are examples of weak adaptations – much more like Kettlewell's moths than entirely new structures in the engineering sense.

Response: Yes, precisely. My claim is exactly that all of the models and evidence involved weak adaptations. But we are still left with a question: Why were these models and evidence *rejected* by most as being evidence for group selection? My suggested answer is: because Maynard Smith, Dawkins, and even Williams (in spite of his originally published view on sex ratios), and so on were looking for strong adaptation, did not find it, and therefore concluded that "group selection" wasn't occurring. These examples of group selection weren't exactly "debated," they were mostly just rejected. I have tried to give an account of why this was.

3. Your Objection: George Williams seems to be looking for something new. In a phone conversation with Wilson he said he was unimpressed with percentage differences such as a female-biased sex ratio that evolve by group selection – he wanted evidence for something more organ-like that evolved by group selection. This is a different position from the Williams of the 1960s, who emphasized the absence of female-biased sex ratios. Now, once these are well established, he seems to require more.

Response: I can certainly understand your frustration at feeling that the opponents to group selection keep moving the goalposts. My insistence on the distinction between engineering adaptations and product-of-selection adaptations was to try to make sense of what, exactly, it was about the models and evidence that the opponents still found unsatisfactory. I think that the best we can hope for is that folks give up using the label "group selection" to refer exclusively to strong group adaptation by group selection – which is how Maynard Smith was using it all along – and start referring to their desired outcome as "strong group adaptation by group selection."

My view about the distinction at the organismic level is this: I have never been primarily interested in the evolution of engineering adaptations, at the organismic, group, or species level. I have argued that focusing on engineering adaptations tends to distract us from other interesting and dynamically significant results of the selection processes (Lloyd and Gould 1993). In fact, the different definitions of adaptation have been around for at least fifty years, perhaps even back to Darwin and Wallace, but nobody distinguished those by giving them separate names until recently. As a result, people have been routinely confused by switching between the engineering and product-of-selection definitions (especially, as I argue, in the units of selection debates.)

I am interested that you perceive that they are using this distinction as a lame retreat. I do not find it lame, since I think the opponents are finally saying what they mean, and are ceding a huge amount of ground on the theoretical and empirical front, that is, the admission that group selection can and does operate in nature, perhaps a great deal. The distinction has enabled them to make an appropriate retreat, which they did not do at all for at least fifteen years, as you know.

4. Your Objection: Group selection has produced adaptations in the strong engineering sense, but it is silly to make the distinction because both strong and weak adaptations at a given level are products of the same process of natural selection at that level.

Response: I think the issues about levels of selection should be primarily about which levels of entities are functioning as interactors. This is the formal definition that we all give. Given this emphasis, the objection is on target: it makes no difference whether the outcomes of a selection process at a given level are strong or weak adaptations; it just matters whether there are entities at that level functioning as interactors in a selection process.

However, other people are interested in other things: some are interested in whether group selection processes can produce engineering group adaptations, and under what conditions, and how often, and what those strong adaptations look like.[1] Given the presence of these other interests, it makes sense to make the distinction, not to give their problems increased visibility or validity, but to make it conceptually possible for them to admit that group selection of the kind we are interested in is present and powerful. From my own experiences in talking with our opponents on this issue, this is exactly what the distinction has succeeded in doing. Moreover, the distinction will allow us to argue about the thing that we genuinely disagree about, that is, whether the focus of investigation ought to be on engineering adaptations at all.

5. *Your Objection:* What is really needed is not the invention of a new conceptual framework for understanding group selection and adaptation, but a consistent application of the old concepts of selection and adaptation.

Response: Yes, of course this is right. But one of my main goals has been to emphasize what had been missed by the other major reviews that were written about your book. Here is a theme repeated in every review I read: why go to these group selection models when we have a perfectly adequate and nongroup selection tools for the job, namely, inclusive fitness models? Why introduce some radical new terminology when our old terminology works well?

The point these opponents failed to get was that their own understanding and application of the old models and terminology is faulty, because they do not actually follow the logic of the models. Those models do *not* require that organismic engineering adaptations evolve, in order to say that there is selection at an organismic level. But these opponents require just that at the group level. Now, because of your occasional appeals to functional integration and complex adaptation, it can look as if you are agreeing with them, that it is reasonable to require engineering

[1] See my discussion of the beneficiary and manifestor-of-adaptation questions (Lloyd 2001).

adaptations at the group level. I think this is part of why the reviewers missed your fundamental point – and an example of a case in which a new conceptual framework really could be used to guard against talking at cross-purposes.

6. *Your Objection:* In arguing against my distinction between the engineering and the product-of-selection views, you note that it is true that Williams does make occasional, obscure remarks about how we can "detect design" by engineering criteria. But you say that these latter remarks are irrelevant to the distinction if they are understood in some loose defeasible sense: the fact that a trait is "complex" and "useful" and "like a machine" are just prima facie indications that it is an adaptation and hence is due to selection.

Response: This is precisely where the distinction pays off. His references to engineering criteria are sincere and contentful. To see why these features are, indeed, indications that something is an engineering adaptation for Williams, consider the conditional associated with the engineering view of adaptation: if something is an adaptation at a level then it evolved by selection at that level. Remember that the logical difference between the engineering and product-of-selection criteria are that the latter and not the former supports the converse: that if something is selected at a level, then it must have an adaptation at that level. Williams's objection to the female-bias sex ratio case turns on precisely his rejection of this converse claim: just because female-bias sex ratios evolved at the group level, that doesn't make them an adaptation for him. His rejection of this converse seems to be clear evidence that he is employing the engineering, not the product-of-selection, criterion of adaptation.

7. *Your Objection:* In *Unto Others*, you endorse what you see as Williams's major contribution to the units of selection debates: that group adaptations are the products of group selection. But, you object, it is wrong to think that adaptations (including group adaptations) must be complicated. Sometimes adaptations involve simplifying a preexisting structure. And it is wrong to think that adaptations have to involve a functional interdependence of an object's parts. Sometimes this is true, but it does not have to be. Hence, my distinction between the engineering and product-of-selection views of adaptation is misdirected.

Response: Does this mean that you endorse that the existence of group-level engineering adaptations can only result from the existence of group-level selection? I don't know anyone who disagrees with this, after Williams made this argument. Or do you mean that anything that results from a genuine group-level selection process is a group-level adaptation in the engineering sense? No, you deny that. So I read this as further evidence of your commitment to a product-of-selection view of adaptation: everything that results from a process of group selection is an adaptation by definition.

8. Your Objection: I wrote in my review that "some of the empirical evidence of group selection that Sober and Wilson cite does not necessarily support the claim that they want to make, i.e., that natural selection has operated on human groups, and that strong adaptations have accumulated at the group level as a result" (1999, p. 448). The best evidence you offer about group adaptations in nature are not about engineering adaptations; these are the cases of female-biased sex ratios, avirulence, and chicken breeding. You say that these are group adaptations (as they are, under a product-of-selection view of adaptation), and you do not think it matters whether they involve "strong engineering design." You want to apply the same standards to group adaptations in humans. The question is simply what traits evolved because of group selection. You neither use nor need a strong notion of adaptation in your book.

Response: Okay, here's how I see it.

1. You want to stick to the product-of-selection definition of adaptation. This is the biconditional version.
2. Therefore, references in your book to complex adaptations, functional integration, organismic-like coordination, are misleading to the extent that they suggest that you are using or have offered evidence for engineering adaptations. One could misread you as endorsing a stronger view of adaptation. I do insist that there is textual support for this; I am aware that Sober's earlier work (1984) endorsed the product-of-selection view, but this book seems to go much further along the road to function, engineering, accumulated, design-type adaptation. At the very least, I conclude that the book is confusing in its presentation of this issue.
3. The evidence: Because of the language mentioned here, I think it is reasonable that some readers will look in your book for evidence of engineering group adaptations. I think they will not find much. Again, I do not consider this a weakness, but others will. As mentioned earlier, all of the cases discussed in the 1970s and 1980s that you cite are *not* cases of engineering group adaptations. That means that the evidence presented in the book will not persuade those, like Maynard Smith, Dawkins, and Trivers, who are looking for evidence of engineering group adaptations. That is the point I am trying to make here.

In sum, I believe that the only way toward any progress on these clash-points is to adhere to consistent and precise terminology regarding what people mean when they connect selection and adaptation. Engineering adaptation implies that selection at that level occurred, but the reverse is not true: selection can occur at a level without producing any engineering adaptation at that level at all. Until this distinction is paid appropriate attention, I believe the units of selection debates cannot be advanced.

ACKNOWLEDGMENTS

I am deeply indebted to Elliott Sober and David Sloan Wilson for their detailed discussions with me over email and in person. I am extremely grateful for their generosity in allowing me to share my version of those exchanges. I would also like to thank Stephen Downes, Herb Gintis, David Hull, Alex Klein, Ken Reisman, Sahotra Sarkar, Robert Skipper, and Hamish Spencer for their very helpful comments on earlier drafts of this piece.

7

Problems with Pluralism

1. SURPRISING ANNOUNCEMENTS

From the 1960s through the early 1990s, substantive and far-reaching changes in population genetics occurred. The debates concern the units of selection, and changed the nature and scope of evolutionary explanations, turning attention from genotypic models to models of structured populations. Furthermore, the relation of a trait to its environment was explored in both models and empirical investigations.[1] For example, Maynard Smith's early theoretical work on group selection provoked theoretical debate, whereas Michael Wade and colleagues produced empirical evidence for the efficacy of group selection as an evolutionary component. Wade's (1978) paper, in particular, criticized ill-conceived, oversold, and rigid theoretical requirements for the efficacy of group selection in nature. The result was a blossoming in a corner of population biology – both in theoretical genetics and empirical studies – of investigations into the possibility of higher-level interactors such as those that had been found in the house mouse by Lewontin and Dunn (1960, 1962) (reviewed in Section 4; for detailed analysis of structured population models, see Lloyd 1988/1994). Various attempts had been made to expand population genetics beyond the organismically focused efforts of the pre-1960s, but the work of the 1980s is distinguished by its achievement of consensus about many fundamentals of hierarchical population genetics and by its empirical substantiation of these ideas. Philosophers of biology soon

[1] Although Goodnight and Stevens (1997) convincingly trace these conflicts of focus back to Sewall Wright, I do not follow the rest of their analysis here.

Parts of this chapter are adaptations of: Lloyd, E. A. (2005). Why the Gene Will Not Return. *Philosophy of Science, 72*, 287–310.

joined the revolution, sorting out the various disputes and definitions, analyzing the variety of evolutionary models, focusing in on what they saw as the key issue of contention: how could the various methods and principles be used for sorting out the key "units of selection" – the objects directly involved in selection processes determining the success of genes, thus engaging in "the interactor debates."

In the midst of this scientific abundance some philosophers, calling themselves "genic pluralists," arose who attacked the new population genetics, its models, and reinterpreted its empirical findings.

In 1990, Philip Kitcher, Kim Sterelny, and Ken Waters (henceforth, "KSW") announced that no one had to concern themselves with the units of selection problem anymore: "Once the possibility of many, equally adequate, representations of evolutionary processes has been recognized, philosophers and biologists can turn their attention to more serious projects than that of quibbling about the real unit of selection." (1990, p. 161). Yet the debates continue. Concerns about units of selection in biology and philosophy have not been settled by claims about representational equivalence.

I argue that three of the chief claims of the "genic pluralists," are defective, and that their overall argument is unsupported. First, there is the earlier-quoted remark about philosophers and biologists who quibble over the units of selection. Second, Sterelny and Kitcher (henceforth, "S&K") also claim that there are no targets of selection (interactors). Finally, KSW claim that they have a concept of genic causation that gives independent genic causal accounts of all selection processes. I argue that each of these claims is either false or misleading. My approach to this literature is unusual; it doesn't take up the proffered notion of genic cause directly, nor does it defend another view of cause. Rather, my approach is indirect, proceeding through a fuller understanding of the role of interactors in selection theory. KSW's claim that debate over units of selection becomes superfluous once the existence of equivalent (read: genic) representations has been established is shown to have things exactly backwards: problems of units of selection must be overcome in order to generate the adequate genic level theory that they take as one of their "many, equally adequate, representations" (KSW 1990, p. 161). Debates over interactors therefore are not "pseudoproblems," and this fact ultimately has fatal consequences for their claims that there are independent genic causal stories for any case of selection (KSW 1990, p. 161).

The burden of proof lies with the pluralists to show that any such genic level causal accounts exist at all. I will not defend any particular hierarchical view in the paper, as these have been defended elsewhere (Lewontin 1970; Wade 1978; Hull 1980; Wimsatt 1980; Sober 1984; Lloyd 1988/1994; Brandon 1990; Williams 1992; Sober and Wilson 1998; Gould 2002; see Section 4).

2. THE GENIC PLURALIST CHALLENGE

"Pluralist genic selectionists" (such as S&K) believe that "there are often alternative, equally adequate representations of selection processes and that for any selection process, there is a maximally adequate representation which attributes *causal efficacy* to genic properties" (1988, p. 358; my emphasis). This sort of pluralism is peculiarly weak: it is simply an equivalence condition. Not only that, the arguments, as given by S&K, entail genic reductionism; an ironic twist, given that pluralism is usually an antireductionist position.[2]

More important, the peculiar sort of pluralism offered by S&K stands in contrast with philosophically significant stronger forms of pluralism. There are three such forms. First, one in which each level of description is understood as indispensable and independent, either locally or globally, for describing a given phenomenon (Dupre 1993). This type of pluralism offers an important type of unity. Each level of description insufficient, and connecting models are needed for the purposes at hand; this type of pluralism can be used to support a global disunity of science claim. A second important and stronger form of pluralism claims that either some or all the different descriptions are jointly relevant to the phenomenon under consideration; this is especially important when the relevant information cuts across different levels or different kinds of description (Darden and Maull 1977; Dupre 1993; Cat 1998, 2000, forthcoming). A third significant type of pluralism concerns unity and pluralism at the level of criteria, independent of any specific theory or application. An example would be multiple causal criteria, wherein criteria of causality above any specific theory contribute a new dimension to the question of unity and pluralism independent of the issue of the connection among theoretical facts (Cat 2000, forthcoming). S&K's pluralism, further details of which are worked out in Section 5, exemplifies the close relationship between pluralism as equivalence and pluralism as reduction, both of which are in tension with

[2] Waters's pluralism, though, is not reductionist, although it does share other weaknesses with S&K's pluralism.

the general treatment of pluralism in the wider literature, as is made clear in the conclusion.

KSW see themselves as attacking "some biologists" who "think... there is a unique account that will identify *the level* of selection" (1990, p. 159; my emphasis). This levels question is the question of identifying the interactors for that process. But, they proclaim, "We believe that asking about the real unit [level] of selection is an exercise in muddled metaphysics" (KSW 1990, p. 159). This is because, according to KSW, the gene itself can always be construed as the interactor in a selection process – the entity that directly interacts with its environment such that replication is differential – once "environment" is construed in terms of the allelic perspective (Waters 1991, pp. 571, 554; S&K 1988, pp. 339, 341, 348). Thus, "[hierarchical selectionists] err... in claiming that selection processes must be described in a particular way, and their error involves them in positing entities, 'targets of selection,' that do not exist" (S&K 1988, p. 359).[3]

Thus, the pluralists' claim that there is a causally adequate, general evolutionary theory purely at the genic level, one that does not require any appeal to higher-level causal interactors. This is not so. By helping themselves to the necessary higher-level (interactor) information, the pluralists make it appear that the hard-won methods for obtaining and incorporating such higher-level causal information (which is represented in the mathematical structure of the models as a whole), are irrelevant, or that debates over these methods have been resolved or overcome by applying the genic approach. They have not. One of the primary claims of my paper is that, contrary to its proponents' claims, the genic account does not give us a theory independent of individuating causal interactions at various levels of the biological hierarchy, nor does it solve or dissolve the problem of how to individuate those very interactions.

3. THE BASICS

In 2001 (see Lloyd 1992, 2001), I analyzed "the" units of selection problem into four distinct questions, and argued that much confusion had arisen because participants in the debates were arguing at cross-purposes.

[3] This may be a spot where Waters disagrees with S&K because Waters does think of selection as a force acting on a target.

The four questions I delineated as distinct "units of selection" questions were: the replicator question; the interactor question; the beneficiary question; and the manifestor of adaptation question. The replicator question concerns which entity passes on its structure directly in replication (usually, but not only, genes) (Dawkins 1982; Hull 1980).[4] I shall expand on the second, interactor question, later, because it is the focus of this paper. The beneficiary question involves which entity benefits, in the long term, from the evolution by selection process, while the question of which entity manifests adaptations plays a central role in determining which entities have "engineering" adaptations at a given level of organization (see Lloyd 2001; Williams 1966). Using this framework, I analyzed many of the major positions in the units of selection controversies, and concluded that numerous players were mixing and matching the various questions in developing their requirements for what it takes to be a real unit of selection.

One of the easy cases for my 2001 paper to sort out was the debate between Dawkins and the group selectionists. Dawkins makes clear that by a "unit of selection" he means a replicator, not a vehicle (interactor).[5] Because he categorizes not only organisms but also groups as vehicles, groups cannot, by definition, be units of selection (1982, p. 115). In contrast, the genic pluralists have developed arguments, driven by new approaches to interactors,[6] which pit genic selectionists directly against group selectionists.[7] Although numerous authors since 1988 have attacked these genetic pluralist views,[8] none of these criticisms has been particularly successful in convincing readers exactly what is wrong with the position (e.g., Sober 1990; Sober and Wilson 1994; Shanahan 1997; Sober and Wilson 1998; Glymour 1999; Van der Steen and Van den Berg 1999; Stanford 2001; Glennan 2002; Wilson 2003).

I argue that the basic problem lies in the pluralists' understandings of the role of interactors in models. Their genic models are explicitly derived from causal models involving higher-level interactors, as I review in Section 5. Any genic causal account is thus derivative from an interactor causal account, and is not independent at the genic level, because it

[4] But see Griesemer (2000).

[5] The distinction between vehicles and interactors is clarified in Section 4.

[6] But see Williams 1966 and Maynard Smith 1987 as underdeveloped precursors.

[7] These are arguments that I did not examine in my 2001 paper, although Lloyd 1988/1994, pp. 133–143 concerns them directly.

[8] Originating in their present form with Waters 1986.

incorporates these higher-level causes. This result undermines their claim to have established independent causal genic selection models at all. Before presenting the details of this argument let me return to the crucial interactor question.

4. INTERACTORS

In its traditional guise, the interactor question asks, what units are being directly selected in a process of natural selection? An interactor may be at any level of biological organization, including group, kin-group, organism, chromosome, gene, or even parts of genes. The interplay between an interactor and its environment is mediated by traits that affect the interactor's effects on genic success. Some portion of the expected fitness of the interactor is directly correlated with the value of the trait in question. Finally, the expected fitness of the interactor is commonly expressed in terms of fitness parameters, that is, in terms of the fitness of replicators; hence, interactor success is most often reflected in and counted through, replicator success.[9]

At what levels of biological organization do interactions occur that make a difference to replicator success? There are a number of ways to study this question, including modeling, experimentation, and fieldwork (see Lloyd 1988/1994 and Sober and Wilson 1998 for relevant literature reviews), and various methods have been proposed over the years for identifying interactors, including the approaches of Lewontin (1970); Sober (1984); Brandon (1982; 1990); Wimsatt (1980; 1981); Lloyd (1988/1994); Sober and Wilson (1994; 1998); and Glennan (2002), among others. These have been used primarily in theoretical and philosophical discussions. At the same time population biologists, sometimes in interaction with philosophers, have developed a wide variety of technical definitions, such as Price (1972); Lande and Arnold (1983); Arnold and Wade

[9] The term "interactor" is David Hull's; it was designed to make up for shortcomings in Dawkins's term "vehicle." Specifically, "vehicle" was meant by Dawkins to refer to the developmental consequences of replicators (not exclusively, but usually, genes). Problems with the vehicle idea occur, however, when genes themselves are the entities interacting with the environment directly, as in meiotic drive or segregation distortion. Using Hull's terminology, genes would, in these cases, be called "interactors," whereas there is no place in Dawkins' hierarchies for them, despite the fact that he emphasizes these "outlaws" from the organismic perspective (Hull 1988a, p. 28; Kawata 1987).

(1984); Arnold and Fristrup (1982); Wade (1985); Heisler and Damuth (1987); Damuth and Heisler (1988); Lewontin and Dunn (1960; 1962); Mayo and Gilinsky (1987); Nunney (1985); and Sober and Wilson (1998), among others. There has been much argumentation over which approach is best (for one superior recent critical evaluation see Okasha 2004). The issue as it stands is undecided, although various research groups using different (but closely related) approaches, each with their various strengths, have made advances.

The emphasis in biological discussions of interactors is on getting the statistical and causal information that will make the model empirically adequate to the phenomena. For example, Heisler and Damuth's popular contextual analysis approach (Multilevel Selection I) to discovering and isolating interactors has three goals: the measurements of relationships between characters and fitness; the location of the *level* of biological organization at which these relationships occur; and the evaluation of *causal models of selection* that are proposed on the basis of prior research (Heisler and Damuth 1987, pp. 594–595). The struggles in the literature involve the best ways of doing so. Some of the empirical work (e.g., house mouse [Lewontin and Dunn], Tribolium [Wade], insects [Colwell], ponds [Wilson], social hymenoptera [Dugatkin and Reeve], crop plants [Griffing], hens [Craig and Muir]) involves demonstrating the efficacy of various hierarchical models in accurately modeling an empirical system (Goodnight and Stevens 1997). Thus, constructing models with information at the higher-level is at the heart of the empirical and causal explanatory uses of hierarchical models. Dugatkin and Reeve emphasize explanatory (1994, pp. 123–124) and research (pp. 126–129) successes of hierarchical models as against "broad individual" models. Sober and Wilson, using the trait group approach to modeling hierarchical selection, review many case studies in which higher-level information makes the difference between building empirically, explanatorily, and causally adequate and inadequate models (1998). Griesemer and Wade (1988) discuss non-trait-group hierarchical models. (See discussion of the differences from trait group models in Wade 1978, 1985; Lloyd 1988/1994; Goodnight and Stevens 1997.) Despite the variety of models and the disagreements regarding the success of particular models, the take-home lesson here is clear. Empirical adequacy, explanatory sufficiency, and research promise are all evaluated to determine the appropriateness of interactor models at distinct levels of the biological hierarchy.

In Hull's analysis, selection must be understood as two distinct *causal* subprocesses: replication and interaction. In the genic selectionism of

KSW (derived, as Waters acknowledges, from Williams 1966),[10] the same processes are involved, it's just that both take place at the genic level. The fundamental claim here is that genes are the ultimate interactors. Whenever other structures can claim to be interactors, genes can claim to be interactors, too, and the genic claims are more "general" and "unified."

I now examine how the pluralist genic selectionists manage to formulate genes as interactors.

5. CLAIMS OF EQUAL REPRESENTATION

The fundamental claim of the pluralists is that anything that a hierarchical selection model can do, a genic selection model can do just as well. Thus, much attention is paid to showing that the two types of models can represent certain patterns of selection equally well, especially those that are conventionally considered hierarchical selection exemplars.

5.1. *Specific Claims of Equivalence*

Let's take a closer look at S&K's example, the sickle cell anemia case, in which the heterozygote is superior to either homozygote in malarial environments. This is usually described as the heterozygotes having a higher fitness than the other pairs, and thus that selection is occurring on the level of the genotype.

Now look at the alleles in the heterozygote superiority case from the genic point of view. "The alleles form parts of one another's environments." "The property of directing the formation of a particular kind of hemoglobin, has a unique environment-dependent effect on survival and reproduction" (S&K 1988, p. 345; see Waters 1991, p. 560). The key lies in considering the other allele at a locus as a crucial part of the focal allele's

[10] Note that Williams does not adopt either Dawkins's later "expansion" of his views, nor Waters's interpretation of his genic line of thought. Williams is committed to using the hierarchical notion of interactor: "Natural selection must always act on physical entities (interactors) that vary in aptitude for reproduction . . . interactors can be selected at levels from molecules to ecosystems, and there has been helpful recent progress on this levels-of-selection question" (1992, p. 38). Although, interestingly, in this 1992 book, wherein Williams spends a good deal of time emphasizing the important causal roles of interactors in selection processes, he still prefers to call replicators (or "codex"), and not interactors, the "unit of selection," because they are the beneficiaries of the long-term selection process (1992, p. 16; cf. Lloyd 2001 on the beneficiary question).

environment, and calculating allelic fitnesses according to these crucial environment parts.

S&K work it out this way in the sickle cell case:

Let P1 be the collection of all those allele copies which occur next to an S allele, and let P2 consist of all those allele copies which occur next to an A allele. Then the property of being the A gene (= property of directing the production of normal hemoglobin) has a positive effect on the production of copies in the next generation in P1, and conversely in P2. "In this way, we are able to *partition* the population and to achieve a Dawkinsian redescription" (1988, p. 347). Thus, an allelic-level description is derived, once the population is divided into the right subenvironments, ones in which the alleles have non-context-sensitive fitness effects.

Note that S&K's redescription requires the values for the positive effects of being an A gene in population P1, and the negative effects of being an A gene in population P2. These numbers are commonly known as the genotype fitnesses. On S&K's story, the fitness of allele A is $W(A) = pw(AA) + qw(AS)$. Thus, they need to have fitness information concerning what is usually considered the (higher-level) interactor in this system, the organism or genotype (AS), in order to derive their Dawkinsian redescription. The reason this is important is because they have offered no principled way of telling when the higher-level causal fitness parameters will be important to their allelic model.

In a paper that some take to support the S&K genic claims, Peter Godfrey-Smith and Richard Lewontin proved that they could create regular population genetics models based on either allelic frequencies or genotypic frequencies (1993). They formalize some of the results implicit in S&K's suggestion for remodeling. The results are quite interesting, because, in Lewontin's ingenious derivation of an allelic from a genotypic model, we find, sitting right in the midst of the "allelic" model, a genotypic fitness parameter. This result (inadvertently) provides further support for the view that empirically adequate allelic level models are dependent *in their construction* on higher-level information. This is a great deal more serious than simply obtaining a parameter value from another model, which happens frequently; rather, S&K have no model at all without relying on the entire higher-level structure.

Waters acknowledges this issue. In his discussion of the Sober and Lewontin (1982) model of heterozygote superiority, he observes that "changes in gene frequencies are determined by the fitnesses of gene pairs... hence, it appears that the force of selection impinges on gene complexes, not on individual alleles" (1991, p. 557). Waters notes that the allelic fitness of gene S with A in its environment may have the same *numerical value* as the fitness of diploid genotype AS, "but the *interpretations* are not the same.... One concerns the propensity of a single gene to make good in a genetic environment, the other concerns the propensity of a gene pair in a less inclusive environment" (1991, p. 560). But what does it signify? If the parameters are semantically distinct and you must use the higher-level information, then the pluralists' models are parasitic, derivative and hence not independent. We still need the information about heterozygote fitnesses, and we need to get it the same way – by looking for interactors in a selection process. Thus, the S&K account is not a genuine alternative to the hierarchical account; it is simply a renaming of parts of the mathematical structure developed through hierarchical means, that is, it is derivative. KSW may object that they are not obliged to furnish new methods or rules of thumb for determining the fitness of a gene. But remember, they are claiming to have a new genic level theory. They have eschewed the units of selection debates, and, presumably, their methods and procedures, having claimed that they are all "muddled metaphysics" and "pseudoproblems." Thus, they cannot just appeal to ordinary interactor-locating devices or methods that arise squarely out of the very debates they have trashed.

Let us now turn to one of the classic cases of the efficacy of interdemic or group selection, the case that even Williams acknowledged was hierarchical selection.[11] Lewontin and Dunn, in investigating the house mouse, found first, that there was segregation distortion, in that well over 80 percent of the sperm from mice heterozygous for the *t* allele also carried the *t* allele, whereas the expected rate would be 50 percent. They also found that male homozygotes with two *t* alleles were sterile. But there was a further complication, which is that, even taking into account the level of the *t* allele segregation distortion, plus the fact that there was strong selection against *t* allele homozygotes, *t* alleles tended to occur at a lower frequency in populations of house mouse than was expected. Consequently, given that they knew the biology of the house mouse tended to favor small breeding groups over large ones, Lewontin and Dunn investigated the effects of differential group extinction. They found a substantial effect of group extinction based on the fact that female mice would often find themselves in groups in which all the males were homozygous for the *t* allele

[11] Note that this is not a case of trait group selection, and, as such, is not subject to any of the mathematical intertranslatability arguments recently highlighted in the philosophical literature.

and hence sterile, and the group itself would therefore go extinct. Thus, they developed a hierarchical selection model on which three levels of interactors were operating simultaneously. This, then, is how a genuine, empirically robust, hierarchical model was developed (Lewontin and Dunn 1960; Lewontin 1962).

What the genic pluralists want to note about this case is very narrow, that is the question "whether there are real examples of processes that can be modeled as group selection can be asked and answered entirely *within the genic* point of view" (KSW 1990, p. 160; my emphasis). Waters tells how to "construct" a genic model of the causes responsible for the frequency of the *t* allele (1991, p. 563).

In order to determine the fitness parameter of a specific allele, let's call it A, we would need to know what kind of environment it is in at the allelic level, for example, if it is paired with a *t* allele. Then we would need information about a further distinct detailed layer of the environment of A, such as what the sex is of the "environment" it is in. If it is in a *t* allele arrangement, and it is also in a male environment, the allelic fitness of A would be changed, as a result of segregation distortion. And so on, with the demic environments, too. As we can see, various aspects of the allele's environment are built up from the gene out, depending on what would finally make a difference to the gene's fitness. The overall fitness of the A allele is calculated by adding up the fitnesses in each set of specialized, detailed environments and weighting them according to the frequency of each environment. Question: How does Waters know that interactions at the group or organismic level will have an effect on genic fitness?

Significantly, each one of these levels of genic environment is an interactor on the ordinary hierarchical view. This is how all the causal information from the regular hierarchical model gets transformed and derived into genic terms. (Note how the allelic model is put together. For each allele, it has its own, nested set of environments: the other allele at the locus, the sex of the gamete, whether or not its companion is a *t* allele, and whether it is in a deme with lots of *t* alleles. Each one of these subenvironments is, on the allelic view, to be considered an environment of the gene, and it has its own fitness.) So the end result is that what was represented as a causally relevant interactor in the hierarchal model ends up being renamed as a causally relevant allelic subenvironment type in the allelic model. *Different state space, same overall fitness structures, same causes.* (See Section 6, Premise 2).

Waters insists that the empirical issues do not disappear under his genic analysis. For instance, the possibility that female mice could be caught in populations in which all males are homozygous for the *t* allele "is as much an issue for the genic selectionist as it is for the group selectionist" (1991, p. 564). Thus, Waters writes, "I can see no basis for concluding that [this genic representation] misrepresents the causal process" (1991, p. 564). "What appears as a *multiple level selection process* (e.g., selection of the t-allele) to those who draw the conceptual divide [between environments] at the traditional level, appears to genic selectionists of Williams's style as *several selection processes* being carried out at the same level within different genetic environments" (1991, p. 571; my emphasis). Note Waters's identification of selection processes with the renaming of parts of the hierarchical mathematical structure (see Section 6).

5.2. How Is It Done?

This whole procedure of determining which level of allelic environment needs to be included looks suspiciously like those used to determine whether something is functioning as a hierarchal interactor. Given that the pluralist genic selectionists have eschewed the interaction question, its presence in the middle of their model is surprising. But they need to know if there are aspects (traits) of the environment functioning at certain levels that make a difference to replicator success. How is that discovered? What are the heuristics for determining relevant partitioning of the environments?

I suggest that what is really going on here is that levels of interaction causally important to the outcome of the genic selection process are being discovered by the usual ways, that is, by using approaches to hierarchical interactors and their environments, such as contextual analysis, and that that exact same information is being translated into talk of the differentiated environments of the genes. For example, take the fact that population content (i.e., how many *t* alleles are in the deme) makes a difference to the selective constituents of that environment. How does the genic selectionist know that the level of population content makes a difference to genic fitness at all? One might say that it is because the division into different clumps of populations at that level leads to the more accurate model. But how do you find that out? The way is just to borrow from the usual ways of isolating hierarchical interactors, and then apply it to an odd way of building models. No independent methods or rules of thumb for determining how to tell how to divide up the evolutionarily significant levels are mentioned for the genic levels.

The similarity between the hierarchical interactor and genic environmental ways of seeking information necessary for an adequate model is clinched by the fact that the pluralists want to use the *same tools* for delineating genetically relevant environments as others do when they are looking for interactors. The point here is not that there is something wrong with the genic selectionists wanting to use efficient tools for dividing up the allelic environments. Rather, the point is that they claimed to have overcome the "quibbles" involving *just those issues*. Waters, however, suggests that a genic analysis could be based on my additivity approach to identifying *interactors*. The additivity criterion presupposes some definition of the environment. Change the way that the environment is defined, for example to Williams's way, suggests Waters, and "genic selectionists could individuate environments" under the additivity approach (1991, p. 563).[12] Unfortunately, the sensible notion of borrowing a method for identifying potential higher-level interactors in order to determine the genic environments and thus to have more adequate genic-level models embroils Waters in the interactor debates that he claimed to avoid using the genic approach.

In a more oblique move, S&K appeal to a traditional approach for identifying interactors in order to divide up genic environments for allelic models in an empirically adequate fashion. Brandon (1982) used the statistical idea of screening off to identify which levels of entities are causally effective in the selection process. In other words, it is a method used to isolate interactors using traditional notions of environments. (Brandon 1990 distinguishes between physical, ecological, and selective environments.) S&K, however, propose that screening off be used for the genic approach by changing the notion of environment to the allelic environment: "Instead, genic selectionists should propose that the probability of an allele's leaving *n* copies of itself should be understood relative

[12] My (1986c; 1988) "additivity criterion" for an interactor has been criticized by Peter Godfrey-Smith (1992) and Sahotra Sarkar (1994). But both Godfrey Smith and Sarkar misread the additivity criterion as requiring that we have a level of selection not only if there is a nonadditive component of fitness at that level, but also if all parameters in the model, including dominance and even the genic level fitness parameter, all have nonzero additivity. This is not what the definition says, as is clear from reading the rest of Chapter 5 (1988/1994). However, I am not currently defending any specific approach to the interactor question. For an insightful account of my additivity criterion, see Griesemer and Wade (1988).

to the total allelic environment, and that the specification of the total environment ensures that there is no screening off of allelic properties by phenotypic properties" (1988, p. 354).

Here, they are discussing using a method – screening off – to determine the correct partitioning of the genic environments that will make an allelic approach work, and that approach was initially introduced into the discussion as an interactor-identifying heuristic. Hence, they seem to be using the same methods for isolating relevant genic-level environments as is usually done for the traditional isolating of interactors. Both can be seen as attempting to get the causal influences on selection right, because they are using the same methods. What is different is that the genic selectionists want to tell the causal story in terms of genes as interactors and their allelic-level environments and not in terms of higher-level interactors and genes. So despite the fact that S&K, like Waters, claim to overcome the units of selection "pseudoproblems," they, like Waters, end up taking sides in the interactor debate.

For Waters and more obliquely S&K, levels of interaction important to the outcome of the selection process (in genic terms) are being discovered in the usual ways – that is, by using hierarchical approaches to identify various levels of interactors – and that information is then being translated into talk of the differentiated and layered environments of the genes.

Given this derivative method of model building, the crucial question for KSW is whether the problem of interactors has really been disposed of. The genic view requires a hierarchical set of environments in order to develop a workable genic fitness parameter. Thus, anything that makes a difference to genic fitness must be partitioned off into a separate environment of the gene. This is done in their examples in only one way – take the already established causally based higher-level model structures, and by terminological transformation and mathematical derivation, "convert" them into genic environments (Godfrey-Smith and Lewontin 1993; S&K 1988; Waters 1991).

Note, in particular, that the relevant *causes* remain at the hierarchical level, for example, Waters's inclusion of demic "environments" as causally relevant. In other words, it is the demes' properties interacting with the demes' environments that are being included here as causally relevant to allelic success; these are read straight off of the hierarchical structures' interpretations of causes. But what has happened here besides a derivative change in state space and the altering of names from "interactor" to "environment part"? Standard hierarchical interactor definitions and techniques are used to isolate, divide, and structure the "environments" causally relevant to genic success, and renaming these higher-level investigations and causes does not make them go away. This is especially important to see in the case of causes; when hierarchical structures' causes are used to produce empirically adequate lower-level models, the fact that the higher-level causes are now "hidden" does not mean they play no role in the new models. More importantly, because the lower-level models are fully derivative from the hierarchical models, there are no *new* causes introduced, there are only causes that are derivative from the hierarchical structures.

The reader may be wondering at this point whether the pluralists' commitment to using interactor methods and models makes any difference to their broader claims, which are ontological. Their exact relations to the alternative account, the hierarchal account, then become the center of attention.

Two models that are mathematically equivalent may be semantically different, that is, they have different interpretations. Such models can be independent from one another or be one derivative from the other. In the genic selection case, the pluralists appear to be claiming that the genic-level models are independent from the hierarchical models. The claim is: although the genic models are mathematically equivalent, they have different parameters, different interpretations, and they are completely independent from hierarchical models.

But, despite the pluralists' repeated claims, we can see *from their own calculations and examples* that theirs are *derivative* models, and thus, that their "genic"-level causes are derivative from and dependent on higher level causes. Their genic-level models depend for their empirical, causal, and explanatory adequacy on entire mathematical structures taken from the hierarchical models and refashioned. This is why their implicit or explicit use of the hierarchical interactor definitions is so damaging to their case that genic models are somehow competitors to hierarchical models.

This point becomes transparent when pluralist genic selectionists use techniques for locating hierarchical interactors in order to make their genic selection models work. By taking a stand on which specific technique to use in determining the levels of causal interaction/allelic environment having significant effects on the genic level, that is, how to subdivide the environments that are taken to be causally relevant to genic success, the pluralist genic selectionists have (inadvertently) taken a stand regarding the very squabbles that they claim to avoid. And they have taken sides in them without undertaking the considerable theoretical work involved in justifying the choices they have made. Thus, they occupy positions that they are unable to defend, due to their denial that there is a problem at all.

Given the failure of the pluralist selectionists to extricate themselves from the interactor debates, and their cavalier and undefended adoption of one or another hierarchical definition of an interactor in order to make their genic models empirically adequate, the punch line to their paper seems especially ill-conceived: "Once the possibility of many, equally adequate representations of evolutionary processes has been recognized, philosophers and biologists can turn their attention to more serious projects than that of quibbling about the real unit of selection" (KSW 1990, p. 161). Instead we have: once the genic selectionists take an undefended position on the units of selection debate, that could allow the possibility of many, equally adequate representations of evolutionary processes, including genic ones. They argued that the presence of equally adequate representations of evolutionary processes means that philosophers and biologists can cease "quibbling" about units of selection.

6. THE GENERAL PLURALIST ARGUMENT

So far, I have focused on how the genic pluralists' models are derivative from higher-level models, in particular their dependence on prior solutions to the units of selection/interactor problem. This might seem to leave their general arguments about the existence of alternate models, and the seemingly arbitrary nature of choice among them, untouched. But this is not so. The issues of derivativeness, dependence, and the general pluralist argument are intimately connected. The basic structure of the argument for pluralism along with evaluations its premises are set out below.

Premise 1: "there are alternative, *maximally adequate* representations of the causal structure of the selection process" (S&K 1988, p. 358, my emphasis; Waters 1991, p. 572).

Analysis:
Derivativeness implies that these alternative representations are not genuine alternatives. Rather they are semantic reinterpretations dependent on the empirical and explanatory adequacy of higher-level models. Investigators into higher-level interactors incorporate environmental factors,

trait-bearing entities and demonstrations of correlations with fitness components, and so on (as reviewed in Section 4) into their models. This results in mathematical structures, (however represented) key aspects of which are required for the empirical adequacy of the structures. Renaming parts of these structures does not change conclusions drawn about other parts of these structures, particularly the level(s) at which entities interact with their environments through their traits (traditionally, the definition of an interactor or unit of selection). Premise 1, therefore, is undermined by the derivative nature of the multienvironmental allelic structures, and especially by the derivative nature of any causal claims. There are no alternative causes, nor operative levels of selection; there are simply renamed structures in derived models.

Subpremise 1: "Specifically, we can always find a way to present a selection process in terms of the *causal efficacy* of genes" (S&K 1988, p. 358; my emphasis).

Analysis:
This premise, a special case of Premise 1, is undermined by Lewontin's argument, Section 8, which shows the dynamical insufficiency of presenting selection processes purely in terms of genic causes in genic state space and reconstructed allelic environments. The premise is already undermined by the derivativeness discussed in the analysis of Premise 1. Lewontin gives an in-principle argument against the possibility of such allelic models.

Premise 2: (Partly suppressed) Genic and hierarchical accounts identify different causal efficacies or agents or causes (S&K 1988, p. 358; Waters 1991, p. 562)

Analysis:
This premise appears to result from metaphysical assumptions involving agents. Specifically, once a model is reformulated in terms of a particular state-space in which alleles are the state variables, there seems to be a tendency to identify a distinct causal agent. An example of this metaphysical move occurs in Waters's account of sickle cell anemia, wherein the "agent" in the model to which we attribute "causal power," as well as a different "level of selection," is the allele, because it is the state space of the model (Waters 1991, p. 564). Waters gives no defense of this position. Williams

defends it by appealing to what I have analyzed as the "beneficiary" question: What entity is around in the evolutionary long run to *benefit* from an evolution by selection process? In sexually reproducing species, the answer must be the allele; hence, according to Williams 1966, the allele is the "unit of selection."

S&K, in contrast, do give arguments for the differences in causal attribution of the hierarchical and genic-level models, but the arguments they appeal to are Dawkins's, nearly all of which are aimed at organismic selectionists, which are not under consideration here (see Lloyd 1988/1994, 2001 for discussion). Dawkins's arguments apply to the replicator question and not to the interactor debate, and hence are misapplied by S&K.[13] Dawkins's own argument is also based fundamentally on his interest in the "units of selection" as beneficiaries, following the lines of Williams, and fares no better against those who are engaged in the interactor debates (Lloyd 2001). The problem with focusing on the "units of selection" as beneficiaries question in the late 1980s is that all of the debate in population genetics, and the vast majority of the debate in philosophy, focused clearly on either the "interactor" question or a combination of the "interactor" question with the "manifestor of adaptation" questions. Thus, KSW, in addressing the "unit of selection quibbles" relating to population genetics, can only plausibly be read as addressing the genetics arguments and related philosophical arguments concerning specific disagreements about how interactors or interactors/manifestors-of-adaptations are defined and identified. Any appeal to the beneficiary question would be, at best, off-point – a metaphysical sideshow.

But perhaps we do not have to appeal to metaphysics at all. There is a precedent in scientific practice for changing causal accounts when the state space is changed. As Rob Cummins has noted (personal communication):

> If you have discriminations that you can make empirically, yet there doesn't seem to be room in your current state space to allow them to do some work, it is a standard practice in the quantitative sciences to move up to a bigger state space, simultaneously changing your interpretation of the model's causal structure. The same thing can work in the downward direction. Thus, we have a principle: change state space: change causal structure.

13 Sterelny and Kitcher repeat Dawkins's arguments that the focus should be on the "causal properties" of alleles, but their argument is undone by their definition of allelic fitness, which falls to Lewontin's objection, below (K&S 1988, p. 346 n. 7).

These cases, however, are purely empirically driven.

With the genic pluralists, in contrast, we change the state space, but not the empirical content of the theory at all, thus undermining the justification for accompanying a change in state space by a change in causal structure. The principle stated above holds where there are changes in empirical content. (Based on discussion with Rob Cummins, January 27, 2005). Thus, the pluralists are trying to have it both ways. An unwary reader, however, may easily be misled by this particular move made by KSW.

Thus, we are back to the metaphysical claims discussed above. In sum, much of the genic pluralists' view rests on implicit, heavily loaded metaphysical assumptions in which causal agency is assigned according to the state space in the model: if you have a model in which the allele is the entity with a fitness parameter, then assume that the allele is the "actor," "agent," or "cause" of the selection scenario.

Conclusion: Genic pluralism is correct (S&K 1988, p. 358):

> "The causes of one and the same selection process can be correctly described by accounts which model selection at different levels" (Waters 1991, p. 572).[14]

Corollaries of Conclusion:

There is no units of selection problem; it has been dissolved by the availability of multiple models.

There are no targets of selection.

Analysis:

The Corollaries of the Conclusion have been shown to be incorrect (see Sections 4 and 5).

Conclusion: Because Premise 1, Subpremise 1, and Premise 2 are undermined, the conclusion is unsupported. In addition, two corollaries of the

[14] This is a legitimate conclusion in the small number of selection cases found that can be categorized as "extended phenotype" cases. But these cases make up a quite small minority of selection cases, and thus cannot be used to represent hierarchical selection as a whole. Dawkins's mistake, in his 1982 book, was to take only organismic selection and genic selection as his two alternatives; thus, he ended up concluding that there was an abundance of extended phenotype cases. Once hierarchical selection is taken into account, this clearly doesn't hold.

conclusion have been shown to be false, which directly indicates the falsity of the conclusion.

7. PRAGMATIC VIRTUES

In presenting their claims for the importance of a completely genic-level evolutionary theory, S&K contrast "two images" of evolution by selection. The two images they have in mind are organismic and genic selectionism. They state that the relative worth of the two images turns on two theoretical claims in evolutionary biology (1988, p. 340):

> 1. Candidate units of selection must have systematic causal consequences. If Xs are selected for, then X must have a systematic effect on its expected representation in future generations.
> And,
> 2. Dawkins's genic selectionism offers "a more general theory of evolution," one that can handle organismic selection cases, but also other cases that are problems for organismic selection.

Challenges to pluralist genic selectionist accounts typically assert that pluralism fails to meet condition 1. Pluralists' arguments purportedly showing the equivalence of genic and higher-level models are aimed at rejecting this critique. As I have shown, this defense comes at a very high price, namely, an explicit appeal (via adoption of hierarchical methods and models) to higher-level causal consequences that the genic pluralists claim to avoid. Nevertheless, the genic selectionists victoriously proclaim: "All selective episodes (or, perhaps, almost all) can be interpreted in terms of genic selection. That is an important fact about natural selection." They continue, "that genic selection could be found in hard cases . . . is surely on any view a major discovery" (KSW 1990, p. 160).

What is claimed to be discovered, exactly? *Mathematical intertranslatability* between various types of genetical models, has already been acknowledged for many years (Cohen and Eschel 1976*; Matessi and Jayakar 1976*; Uyenoyama and Feldman 1980*; Matessi and Karlin 1984; Wilson 1975*, 1980*; Alexander and Borgia 1978*; Colwell 1981*; Crow and Aoki 1982*; Michod 1982*; Wade 1985*; Grafen 1984*; Ratniecks and Reeve 1991; this list of citations is compiled from Kerr and Godfrey-Smith 2002 and Dugatkin and Reeve 1994; asterisked papers are discussed at length in Lloyd 1988; see also discussion in Sober 1984;

Sober 1990; Sober and Wilson 1998). The presence of these models has received a recent boost in visibility in the philosophical literature from the work of Dugatkin and Reeve (1994) Sterelny (1996) Kerr and Godfrey-Smith (2002), and Godfrey-Smith and Kerr (2002), who have emphasized the intertranslatability between some sorts of group and structured population individual models, respectively. Dugatkin and Reeve's claimed equivalence between the above cited models and an allelic causal model is false (Lloyd, Lewontin, and Feldman, 2006). In any case, the Godfrey-Smith and Kerr findings won't really do much for the problem facing the genic selectionists, since all of their lower-level models help themselves to *higher-level fitness parameters* in order to make them work, without any principled or theoretical justification, which would be beside the point for their particular purposes.

Much more substantively, there is the supposed discovery by the pluralists that *there are no targets of selection (interactors)*. But we see a tension between their methods and their conclusions. "Empirical questions about the existence of group selection are reformulated by genic selectionists in terms of the existence of multiple selection processes occurring in different genetic environments" (Waters 1991, p. 572). They use ordinary hierarchical techniques to determine the appropriate levels of mathematical representation of the suitably partitioned, allelicly sensitive environments of the alleles, then return to say that the methods they've just used are superfluous, and that people shouldn't argue about the interactor question at all. This approach has all the advantages of theft over honest toil. The "targets of selection" are still there in the models, they are simply renamed. By what sort of scientific metaphysics is renaming enough to make something disappear entirely?

Sterelny and Kitcher refer to Dawkins's defense of the view that "genic selectionism offers a more general and unified picture of selective processes than can be had from its alternatives" (1988, p. 354). It is important to note, here, that by "its alternatives," Sterelny and Kitcher are referring explicitly throughout to organismic selectionism, and not to the actual picture held by the opponents that they take on, namely, hierarchical selectionism, except in two of the last three pages of the paper. This is an extremely odd feature of the Sterelny and Kitcher article.[15] The article is taken up with criticizing the views of Stephen Jay Gould, Richard Lewontin, Elliott Sober, and Robert Brandon. All of these authors are well established as defending a hierarchical selectionist view; that is,

[15] Waters's 1991 paper does not suffer from the same fault.

they believe that it makes sense to look for interactors at various levels of the biological hierarchy, from the gene to organism to the group or above. They most definitively do *not* think that selection theory should be approached as if organisms were the only interactors. Nevertheless, nearly all of Sterelny and Kitcher's arguments are aimed at those who take an organismic approach, rather than a hierarchical one. On the face of it, the hierarchical approach to selection is a serious contender for a good, general theoretical approach to evolution by selection. It takes easy account of the "outlaw" examples of competition among genes that Dawkins (and Sterelny and Kitcher) trumpet as problems for the "alternative accounts."[16] (And, in fact, it is one of the originators of one of the best empirical cases of an "outlaw," the *t* allele.) It takes extended phenotype cases on an ad hoc basis just as the genic approach does. Thus, there is no sense in which the genic approach is obviously superior in generality to the approach already in place and promulgated by those criticized by Sterelny and Kitcher.

When Sterelny and Kitcher do finally get around to confronting the real contender, hierarchal selection theory, they seem to misunderstand it. The difference, they claim, between hierarchical monism and pluralist genic selectionism is that "whereas the pluralist insists that, for any process, there are many adequate representations, one of which will always be a genic representation, the hierarchal monist maintains that for each process there is just one kind of adequate representation, but that processes are diverse in the kinds of representation they demand" (1988, p. 359). What Sterelny and Kitcher have failed to appreciate is the unity of the hierarchical view: its interactors and replicators all the way up and down the biological hierarchy, each playing their causal roles in the selection process (more on this in a moment).

In addition to the supposed haphazard diversity in the kinds of representation demanded by the hierarchical view, it also has against it that it is "entangled in spider webs," that is, in extended phenotype cases (1988, p. 360). It seems that Dawkins might have changed his mind about extended phenotypes since the old days. In 1994, he praises interactors because they are much better at modeling extended phenotype cases than are his usual concepts of "vehicles." "Darwinism can work on replicators

[16] In fact, contra Sterelny and Dawkins, Dugatkin and Reeve argue that a two-level model of selection is needed to model intragenomic conflict such as meiotic drive (1994, p. 124). This is typically how meiotic drive has been modeled in hierarchical selection theory.

whose phenotypic effects (interactors) are too diffuse, too multi-leveled, too incoherent to deserve the accolade of vehicle" (1994, p. 617). This is in tension with Sterelny and Kitcher's claim that, for extended phenotype cases, "there are genic representations of selection processes which can be no more adequately illustrated from alternative perspectives" (1988, p. 360). In addition, I would point out that hierarchical selectionists are free to help themselves to any genic-level model representations that seem handy or the most appropriate. Thus, the extended phenotype has no bite against the hierarchical view, only against the organismic selection view, a logical point that Sterelny and Kitcher miss.

At any rate, according to Sterelny and Kitcher, genic selectionism is still the most general approach, and has nothing to fear from hierarchical selection theory. Going back a bit, this weird disconnect between Sterelny and Kitcher's primary target – organismic selection – and their real opponents – hierarchical selectionists – was noted immediately by Sober in his 1990 commentary on their 1988 piece (p. 152). Sober went on to note that the units of selection problem was not actually a notation or vocabulary problem, and did not consist in whether or not it was possible to *represent* a system in one form or another, but rather consisted in empirical questions concerning biological processes. He then spelled out a case in which group selection was discovered, trying to demonstrate the sort of issue that was at stake in the interactor debates. Kitcher, Sterelny, and Waters acknowledge that Sober's case would indeed be a real discovery, but then dismiss it in two ways, first by noting that it does not vindicate Wynne-Edwards style group selection (an irrelevant point), and second by claiming that the "discovery" of a process of group selection "does not invalidate individual-level or genic-level descriptions" (1990, p. 159). In other words, the discovery of a group-level process does not establish that there is a "uniquely correct" description of the causal process, one that rules out a genic description of *that same process*. This interchange could hardly make clearer that Kitcher, Sterelny, and Waters have attempted to change the subject from the interactor debate; they have not answered it, or dissolved it, as they flatly assert. And they have every right to change the subject. But what is remarkable about their move is that they seem to think that by attempting to change the subject – to one of representation – this gives them the right to draw conclusions about a debate they have claimed to overcome and shown as trifling, but have become more-or-less covert and deep participants in, the interactor debate, without having done any of the hard work.

8. CONCLUSION

In sum, the genic view does not have the big advantage claimed; it does not avoid the unit of selection "quibbles" after all. But the causal questions concerning which entities are interacting with their environments in a way that affects replicator success – including questions concerning how to know when to take such interactors into account and how to detect them – are crucial for the genic approach, as they themselves have shown. The causal equivalence of the models is better thought of as derivative; higher-level causes are reformulated down in model translation and imported into the genic models, not changing one whit the level at which the identified interaction between trait and environment that makes the difference to genic success (the interactor) actually appears in the overall mathematical structure from which the genic model is derived.

This means that the crucial question regarding the levels of interaction necessary to model a given selection process is still outstanding, whether it is put in hierarchical terms or genic ones. Worse, the pluralists we examined attempt to sidestep this question, but cannot; they end up taking sides inadvertently in the interactor debate, all the while claiming they have overcome just that debate. Neither philosophers nor biologists can ignore the debate about interactors in selection; the outcomes make a difference to the empirical, explanatory, and causal success of the models produced.

Even if we wholeheartedly endorsed the genic selectionist desire to make genes both the replicators and "real" interactors in the evolutionary process, some formal equivalent to the interactor question would have to be answered in order to make the genic approach empirically adequate, not to mention causally adequate, as they claim it already is. They need some way to divide up the causally relevant allele sub-environments, and thus far, have only borrowed from the hierarchical theory, all the while trashing it. This demonstrates, I think, a rather profound failure on the part of the pluralists to come through with a genuine autonomous genic theory, as well as a failure to understand what is required from a selection theory overall.

Pluralist genic selectionists also claim that, once genic and other alternative representations of selection processes are available, there is no scientific way (except perhaps the reductionist one) to choose among the various alternative models (S&K 1988, p. 358).[17] In addition to the

[17] Waters (1991) is probably an exception to the reductionist move, here.

derivativeness argument, the additional flaw with this argument is that Lewontin has shown the dynamical insufficiency of their approach to genic models, wherein the allelic frequencies and the layered allelic environments of the alleles determine fitnesses. Even if they had an independent causal definition at the genic level (which they do not), the models would not be the promised "[presentations of] a selection process in terms of the causal efficacy of genes" (S&K 1988, p. 358). Thus there can be no "causal equivalences" of the sort they claim to have shown.

Let me spell this out in more detail.

Perhaps S&K, especially, have simply chosen badly in illustrating their point, and perhaps the revamping of Williams that Waters is suggesting could do the trick without relying on the derivation from hierarchical models. Not so, says Lewontin. There is something fatally wrong with any model that attempts to model a selection process using genic selection parameters and associated allelic environments. Indeed, there is also something faulty in using purely genotypic state space models to represent selection, since, at every generation, we need allelic-level information about segregation and, usually, recombination, in order to produce a pool of gametes. Thus, properly conceived, population genetics models need, at minimum, information from both the allelic and from the genotypic space. Because all regular population genetic models require both allelic- and genotypic-level models to produce dynamically workable models, Lewontin argues the following. Take an allelic-level model in which we are using allelic-environments to determine allelic-level frequencies (as all the models suggested by the pluralists require). At each generation, we must readjust the selection coefficient according to the frequencies of the different alleles. How is this done? The obvious way is to take a weighted average of the fitnesses in the different allelic "environments" weighted by how frequent the different "environments" are (as suggested by S&K 1988, p. 346 n. 7). But this will only work if the alleles find themselves in the different environments randomly, that is, if we are in Hardy Weinberg proportions. So we are required to determine how frequent the combinations of alleles and "environments" are, that is, what the frequencies of different diploid genotypes are. (That is, we are required to go up to the higher level: it cannot all be done at the allelic level, any more than genotypic calculations can all be done at the genotypic level alone, but are reliant on the allelic level.) Even if the population is in Hardy Weinberg equilibrium, that must be confirmed, which also requires going up to the genotype level. So, no matter what, in order to calculate the next generation of allele frequencies, we must know the next generation's genotype frequencies, in order to calculate the fitnesses of the alleles in the allelic space. The only alternative is to stay in the allelic space completely, and track empirically the changes in the genic selection coefficients at each generation, in which case, every generation would produce a new model, unlinked to the model of the previous generation. If we were to approach it this way, there would be no science of population genetics, that is, no way to produce trajectories of gene changes over time, or predictions of changes, or explanations of changes. Consideration of more than one locus for selection (as many of the models touted by the genic selectionists involve) makes the empirical tracking approach mathematically intractable, as well as predicatively useless. Thus, *there is no way to actually represent a selection process over time with an allelic-level model.* They are not dynamically adequate without information from at least the genotypic level, and maybe higher levels, so even setting aside the issue of the origins of the allelic models from hierarchical or genotypic models, the allelic models cannot be used at all without the constant input of genotypic or hierarchical models in order to represent selection processes. There is no equivalence here, only utter dependence. (Based on personal correspondence with Richard Lewontin, January 15, 2005.)

Obviously, this argument completely destroys Subpremise 1. But even if Lewontin's argument were to fail, and through some other form of modeling, an independent purely genic/environmental causal model could be rendered empirically adequate, KSW have certainly not established this. Meanwhile, there are no "causal equivalences," and so no independent genic models. Thus, there are no independent maximally adequate causally different accounts of any particular selection episode to chose between: there are no independent genic causal models, and all the other models offered as "alternative views" are derivative. When combined with the problems with Premise 2, the genic pluralists are left in trouble.

Not only have the genic pluralists failed to show that the units of selection question is obsolete, they have failed to show their famous "equivalence" between alternate causal models of selection, particularly the genic ones: generally, such the supposed "alternative causal models of selection are (1) mere derivations of hierarchically structured models, and not causal alternatives, as advertised, and (2) Lewontin shows that, without such derivations, the genic models are dynamically inadequate.

The only sustainable conclusion of "The Return of the Gene" and "Tempered Realism" turns out to be that accurate, causal, empirically adequate genetics models are precisely the hierarchical structures that KSW have relabeled.

Thus, rather than the very weak (and unworkable) equivalence form of pluralism KSW find in the units debates,[18] I see four factors pointing toward one of the stronger forms of pluralism mentioned in Section 2: the *t* allele case; the methods for isolating and identifying interactors; proven derivativeness; and Lewontin's argument for the unsustainability of an allelic-level model-type. In particular, descriptions of some or all of the different levels of a system are required for an adequate scientific account (see Section 2). This hierarchical form of pluralism instantiates locally, for a given phenomenon, a non-reductive form of unity, because all the different forms of descriptions are brought to bear. Thus, hierarchical selection exemplifies a strong and important form of pluralism, exactly contrary to the claims made by KSW; far from being a form of "monism," as charged, it is a strong form of pluralism, standing in contrast to the weakened (and reductionist, in the hands of S&K) form of pluralism advocated by the genic pluralists.

Sometimes simpler models are useful as tools, as has long been recognized by all geneticists (for recent discussion, see Michod 1999). But playing with the genetics in such a way that it doesn't improve the empirical adequacy of the models or theoretical understanding of evolutionary change is simply a step backwards to 1966.[19] Remember, Williams's (1966) allelic parsimony argument was widely interpreted as a reason to never even take a look at higher-level interactions, interactions that eventually led to empirically more adequate models (Lloyd 1988/1994, pp. 94–96; Williams 1966, pp. 5, 66, 93–95; Williams 1992). Similarly, the genetic pluralists seem to want to return us to a pre-1980s state of population genetics, one in which structured population genetics and hierarchical selection structures played little or no role, empirically or theoretically.

In the end, confused metaphysics (or twisted scientific practice, take your pick) are insufficient to fill the gaps in the arguments the genic pluralists present. Thus, genic pluralism fails on multiple counts; it serves only to highlight the success of hierarchical approaches to selection processes.

[18] And the genic reductionism of K&S.

[19] Kerr and Godfrey-Smith's (2002) recent article claims to do more than this; specifically, they believe viewing an evolutionary process from both individual and trait group point of view yields insight that might not otherwise be available. Note that their paper helps itself to the higher-level information from the beginning, clearly begging an important empirical and methodological question. Nevertheless, it is also heavily indebted to Eshel's work from the 1970s, work that lies firmly in the tradition discussed in the present paper, and so might well be compatible with many of the conclusions drawn here.

8

Normality and Variation

The Human Genome Project and the Ideal Human Type

Certain issues involving science are widely regarded as ethical or social – the appropriate moral and medical responses to abnormal fetuses, for example. The "concept of abnormality" *itself* is not usually one of these social or ethical issues. It is assumed that science tells us what is normal or abnormal, diseased or healthy, and that the social and moral issues begin where the science leaves off.

For many purposes, such an understanding of science is appropriate. In this chapter, however, I would like to challenge the "givenness" of the categories of normality, health, and disease. By understanding the differences among various biological theories and the distinguishing features of their respective goals and approaches to explanation, we can analyze the way in which scientifically and socially controversial views are sometimes hidden inside apparently pure scientific judgments.

I am not suggesting that the misleading nature of some of the scientific conclusions I discuss implies some unsavory *intention* on the part of the scientists involved. On the contrary, my point is that there are sincere scientists working among different theories and subfields, each with their own standards of explanation and evidence. The diversity of theories and models involved in implementing the Human Genome Project provides a unique challenge both to the producers and the consumers of the DNA-sequencing information. I contend that the problems arising from this diversity have not been recognized or addressed.

In drawing attention to some important differences among the biological theories involved in the Human Genome Project, I will explore several

Lloyd, E. A. (1994). Normality and Variation: The Human Genome Project and the Ideal Human Type. In C. F. Cranor (ed.), *Are Genes Us? The Social Consequences of the New Genetics* (pp. 99–112). New Brunswick, NJ: Rutgers University Press.

possible misunderstandings that may be arising from viewing biology as a monolithic and completely integrated science. Furthermore, I raise some concerns about the risks inherent in biological and medical reasoning about genetics.

HEALTH AND DISEASE

Judgments concerning health and disease inevitably involve questions of classification. "Health" encompasses the thriving, fully functioning, or normal states of the organism, whereas "disease" includes states of malfunction, disturbance, and abnormality. States of organisms do not announce themselves as desirable or undesirable, healthy or diseased, normal or abnormal; such classifications are inevitably applied by comparing the state of the organism to some *ideal* which serves a normative function. Where does this ideal come from?

Roughly speaking, our notions of the ideal state of an organism are informed by our understanding of the "proper" or appropriate functions of various parts. The function of the kidneys is to clean the blood; if the blood is not cleaned thoroughly, and the organism loses the benefits of the "proper functioning" of the kidneys, then the kidneys are "diseased." Overall, the organism does not function as well as it once did. But suppose instead that an organism with badly functioning kidneys never had kidneys that effectively cleaned the blood? Then the comparison must be made not to the prior state of that individual, but to a more abstract notion of "proper functioning" of kidneys in people. In other words, the ideal is *normal* kidney functioning, where "normal" signifies the function in a thriving person.

The difficulties of classifying diseases are well known, having been faced by every theory of medicine in human history. The range of definitions of "normal functioning" – from proper balance of the four humors, to clear flow along the chi meridians, to freedom from cohabitation with microorganisms – has also received a great deal of attention. It seems, therefore, that it would be a major scientific advance, and a significant relief, to be able to understand disease and proper function on the molecular level, and it is just this that is promised by many proponents of the Human Genome Project.

Renato Dulbecco, for example, argued that significant advances in cancer research would be made possible by knowledge of the exact human DNA sequences (Dulbecco 1986, p. 261). Similarly, Nobel laureate Jean

Dausset has defended the Human Genome Project by arguing that it will allow us to predict when a person will develop an illness, or at least when he or she will have a predisposition to that illness. Early diagnosis and preventive measures are also emphasized. Dr. Jerome Rotter, director of the Cedars – Sinai Disease Genetic Risk Assessment Center in Los Angeles, says, "[I]t's more than just knowing you're at risk. You can take steps to prevent coming down with the disease or be able to cure it at an early stage" (Roan 1991).

The descriptions of disease promised by the Human Genome Project are intended to be on the biochemical and, sometimes, even the molecular level. James Watson, codiscoverer of the structure of DNA, and later director of the National Institutes of Health segment of the Human Genome Project, asserted that genetic messages in DNA "will not only help us understand how we function as healthy human beings, but will also explain, at the chemical level, the role of genetic factors in a multitude of diseases, such as cancer, Alzheimer's disease, and schizophrenia" (Watson 1990, p. 44). The geneticist Theodore Friedmann has proclaimed that, "molecular genetics is providing tools for an unprecedented new approach to disease treatment through an attack directly on mutant genes" (Friedman 1989, p. 1275). Once diseases have been pinpointed on the molecular level, treatment can begin: "inherited diseases can be identified with biochemical as well as genetic precision, often detected in utero, and, in some cases, they can be treated effectively" (Culliton 1990, p. 211; see also Jukes, 1988).

The promise is that diseases will finally be subject to truly scientific classification, analysis, and treatment. Detailed descriptions of the molecular causes of disease will enable medical researchers to develop more precise preventive and therapeutic techniques. Once the human genome is sequenced, we will have a library of genes with which any potentially abnormal gene can be compared. Abnormal genes will be isolated, altered, replaced, or, in case they are present in implantable embryos, simply discarded.

In many ways, this picture of medical promise and possibility is undoubtedly positive. It is also inadequate and misleading. The primary problem is that the picture of disease most often presented in discussions of the Human Genome Project is oversimplified. Specifically, the presentation of genetic disease and abnormal gene-function as self-announcing is unjustified, except in the most trivial sense. General physiological notions of normality, health, and disease are defined according to a *different* set of standards that go beyond molecular-level descriptions. Describing genes as "causing" diseases is, on a basic scientific level, to confuse at least two distinct levels of theory and description. Although a genetic classification of disease may indeed be desirable and useful, it also involves a series of judgments about the ideal forms of human life. Moving the level of diagnosis down to the molecular level does *not* succeed in avoiding the fundamental value judgments involved in defining health and disease, contrary to the suggestions of the genome researchers.[1]

So although molecular techniques will certainly aid in the diagnosis, identification, and analysis of disease processes, they cannot replace the profoundly evaluative and essentially social decisions made in medicine about standards of health and disease. In fact, molecular techniques should be understood as offering an unprecedented amount of social power to label persons as diseased. Hence, it is more important than ever to gain insight into the normative components of judgments about health. The potential submersion of normative judgments under seas of DNA-sequence data should not persuade anyone that conclusions concerning health and disease have now, finally, become scientific. Any appearance to the contrary is a result, I will argue, of some rapid and illegitimate shifting between biological subtheories. Once the structure of the theories involved has become clear, it will be easier to see how and where evaluative decisions are being made.

MOLECULAR DESCRIPTIONS

Let us begin by taking a closer look at the description and explanation of disease at the DNA level. The typical form of the model used in explanation is described by T. H. Jukes: "[I]nherited defects would be caused by changes in the sequence of DNA, perhaps a change in a single nucleotide. Such change might result in the replacement of one amino acid by another in a protein at a critical location, making the protein biologically useless" (Jukes 1988, p. 16).[2]

[1] Another misleading aspect of the typical genetic presentations of disease concern implicit assumptions of genetic determinism and disease. See the following for relevant discussion: Karjala 1992; Gollin, Stahl, and Morgan 1989.

[2] Friedman describes the goals of molecular genetic explanations also: "Predicting the exact structure and regulated expression of any gene, the tertiary and quaternary structures of its products, their interactions with other molecules, and finally, their exact functions, constitutes a problem with the highest priority in molecular genetics" (1990, p. 409).

A paradigmatic case of genetic disease, sickle cell anemia, fits this general model well and is an early, compelling, and fairly complete case of medical genetics. The original mystery was why certain populations had a high incidence of a type of red blood cell that seems to cause health problems. The red-cell abnormality stemmed from a difference in the hemoglobin molecule which impeded its ability to carry oxygen, hence damaging the bodily capacities of a person with these cells. An analysis of the biochemical causal pathway revealed that the genes that code for the hemoglobin molecule in these people were different from other people's hemoglobin genes in a particular way, thus affecting the ability of the hemoglobin to pick up oxygen. The differences in genes contributed to a difference in the proteins made according to the pattern on those genes, and these protein (hemoglobin) differences had systemic and detrimental effects on health.

This *biochemical causal-pathway model* traces an isolated chain of events that yields the effect of interest. In the standard genetic/biochemical model for the production of hemoglobin, the hemoglobin gene codes for a protein that is included into the red blood cells, cells that, in turn, serve the function of carrying oxygen to the cells of the body. The model presents, in detail, a picture of "normal or proper functioning."

There is not much room for simple variation in this explanatory scheme using the standard model.[3] But why should there be? The purpose of the basic explanation is to explain *how it could be* that hemoglobin is produced and operates effectively in the body. The goal of the explanatory theory is to delineate at least one causal chain that could proceed from the initial state – DNA arranged on chromosomes in a zygote – to the final state – iron arranged in hemoglobin molecules carrying oxygen around the body. The final model of hemoglobin abnormality in sickle cell anemia represents an astonishing piece of detective work, because it assumes an understanding of each chemical reaction involved in the "normal" causal chain. The abnormality itself is explained through isolating the points in the sickle cell causal chain that are different from the model of normal functioning.

Watson, in motivating the genome projects, presents a similar picture: "The working out of a bacterial genome will let us know for the first time the total set of proteins needed for a single cell to grow and multiply"

[3] James Watson speaks ambiguously of the "genetic diseases that result from *variations* in our genetic messages" (1990, p. 46, emphasis added). Not all variations result in genetic diseases.

(1990, p. 48). Such a goal is perfect for a biochemical causal model: what is desired is some complete set of causal steps yielding a living organism. But variation plays no role in this model. It is an uninteresting, and even distracting, feature of the processes on which the explanatory theory is focused.

In other words, under a biochemical causal model, there is no obvious approach to dealing with variation. One could classify all variations on this main scheme as "abnormal." Owsei Temkin, in ridiculing a definition of disease based on genetic origins, teased that "there should also be as many hereditary diseases as there are different genes representing abnormal sub-molecular chemical structures" (Temkin 1977, p. 444). Temkin argues against classifying diseases on causative principles in general, "lest specific diseases be postulated which have no clinical reality" (1977, p. 444). More sensibly, we would prefer to define a variant as abnormal or diseased only if it interfered with "proper function." Proper function, however, cannot be defined within the molecular genetic model itself: it must be defined in terms of the physiological functions of the resultant protein.[4] Here, proper function could involve simply the presence of an *effective* hemoglobin for carrying oxygen around. But there could be (and are) different degrees of effectiveness. How should we divide these up into abnormal and normal?[5] Simple variations and undesirable variations? Some differences in DNA lead to large changes in function, whereas other differences are imperceptible with regard to oxygen delivery in a normal person.

A friend was exposed to this problem personally. Having gone to a physician who was up-to-date on all the most recent screening techniques, my friend's slightly red eyes prompted the doctor to run a blood test. The results of this test were positive, and my friend was informed by his doctor that he had "abnormal" hemoglobin and liver functioning, something called "Gilbert's disease." When my friend asked about the health consequences of this abnormality involving an essential protein in his body, he was told that the disease was "nonfunctional," and the only known effect was a reddening of the whites of the eyes.

[4] Gollin et al. include in their list of uses "an index of acceptable physiological function or behavior" (1989, p. 50). Such views derive ultimately from Herman Boerhaave, who defined diseases as occurring when a person could not exercise a function.

[5] Theodore Friedman admits that a great deal of sequence information and "comparison" is needed in order to distinguish between "polymorphisms" and "disease-related mutations" (1990, p. 409).

The point is this. If normality is defined at the level of the biochemical causal model, all variation in the DNA of the hemoglobin genes is abnormal. As such, however, the genetic abnormality tells us nothing about its effects on physiological function in the larger organism.

Another example emerges from recent research on human breast tissue. "Fibrocystic breast disease" exists in approximately 45 percent of the female population over age thirty-five. In one sense, the relevant phenomena are called "fibrocystic disease" because they involve the process of encystation, which is additional to the usual physiological *functions* of breast tissue, such as milk production. On the medical and physiological level, however, it seems that the fibrocystic condition is completely "normal" for the average adult woman; furthermore, the condition does not impede or alter the usual physiological functions of the breasts. In what sense, then, is the fibrocystic condition a "disease"? This is a case in which refined understanding at the cellular level leads to the classification of a condition as an "abnormality" or "disease," whereas at the functional, medical, or physiological level, it is unclear what sense can be made of labeling nearly half the women over the age of thirty-five "diseased."

Again, if normality is defined according to some model of physiological function, molecular information alone cannot decide whether a certain person is normal or abnormal. The DNA information itself is potentially revealing about the functional state, but only potentially.[6] Abnormality in the DNA or in the causal chain may or may not have health consequences that we would consider significant. Carl Cranor, in his discussion of genetic causation and Hartnup disorder, emphasizes the contingency of the emergence of disease on other factors (see Cranor 1994). He also reviews the various types of confusion that can arise regarding genetic causation. I offer here a diagnosis of the underlying mechanism that produces the kinds of confusion discussed by Cranor.

PROPER FUNCTION

The second type of biological model involves reference to proper function on an organismic level. This level of model is usually considered most appropriate to medicine, and it could be called a *medical model*. Generally speaking, the medical model tends to be on the level of the whole organism, rather than on the cellular or molecular level. Take the example of the

[6] See Friedman's (1989) discussion of "normal, functional" genes.

common cold. Although part of the explanation of how a person contracts a cold is that they were exposed to a cold virus, the rest of the explanation requires taking account of the body as a whole: one does not get a cold simply from exposure to the virus; failure of the immune system to fight the virus invasion effectively is necessary, as is multiplication of the virus within the cells. Similarly, cancer cells are recognized as such on the microscopic level, but their undesirability is because of the damage they do to organ function. The presence of cells simply growing in the wrong place may not impair function – witness the innumerable cases of benign tumors that lie undetected until death occurs from other causes.

The first important aspect of the medical model, then, is its *organismic basis*. A second feature, one that has far-reaching social and policy implications, is that the medical model must rely on *socially negotiated standards* of what counts as the proper functioning of a human being (Temkin 1997, p. 447; cf. Caplan 1989). What range of functional performance is normal? Any answer involves a picture of what a human body should be like.[7] Probably the clearest recent demonstration of the social negotiation of categories of health and disease is the battle, in the past two decades, over the medical classification of homosexuality (Bayer 1981). Peter Sedgwick (1973), for example, argued persuasively that classifying homosexuality as a disease is clearly not just an empirical assessment of biological function.

To see the force of this argument, take the claims made in 1991 concerning differences in brain structure between homosexual and heterosexual men (Levay 1991). Suppose, for the sake of argument, that these anatomical differences arise (in this environment) from genetic differences. (There is no evidence for this; the differences could just as well be caused *by* homosexual activity as be the causes of it.) Suppose, furthermore, that we were able to isolate some genes whose functions included structuring this part of the brain, and that people with a particular sort of brain structure were more likely to be homosexual. Should this genetic character be considered an "abnormality"?

To answer the question, we would need to assume some "proper functioning" of the brain structure, and we must also align this proper function with a particular environment. (It could be that the same genes produce a different brain structure in a different environment, and that this brain structure is not correlated with a tendency to practice homosexuality.)

[7] Caplan (1994) offers a useful summary of the debates about the extent of values in definitions in health and disease.

What can we conclude about these genes? Only that they yield particular results in particular environments. But is this normal functioning? Clearly, the answer depends on something *outside the genetic causal story*: it depends on whether we think homosexual behavior should count as normal functioning in human beings. In other words, this distinction between normality and disease depends on how we envision human life ought to be. This story introduces a further problem. If homosexuality is not seen as normal functioning, it is unclear which approach would most successfully move the population toward a higher incidence of normal functioning – changing the environment in which this gene is expressed, or doing something on the genetic level to select out or replace this gene.[8]

The example of homosexuality brings out the profound value decisions involved in labeling certain functions as normal or abnormal. Without such evaluations, genes cannot be labeled as normal or abnormal in any but the most trivial respect – that is, insofar as they differ from the paradigmatic biochemical causal pathway currently accepted for that gene.[9] And such a weak classification system cannot do the work in medical genetics that has been advertised for the Human Genome Project.[10] The issue of defining the *standard* of health and disease is as open as it has ever been. Indeed, there is now the additional challenge of applying it to unimaginably fine biological differences.

POPULATION GENETICS MODELS

Very different types of description and explanation are used in population genetics, where the emphasis is on the analysis and maintenance of variation in populations. Population geneticists have posed persistent challenges to the grander claims made for the Human Genome Project, urging that a biological account of variation in human populations must

[8] See the discussion of Cranor (1994) and in Hubbard (1990).

[9] Gollin et al. describe this unjustified shift from developmental theory to the physiological level as follows: "The observed regularities [in development] are mistaken for universals and the construed universals are regarded as indicators of health and developmental adequacy" (1989, p. 51).

[10] Even Theodore Friedman, who is committed to the medical benefits of the Human Genome Project, discusses the distance between DNA-level characterization of an "abnormal" gene and a medical and physiological understanding of the disease. "Most successes of medical genetics have begun with an understanding of an aberrant metabolic pathway; the genes responsible were identified and isolated through this physiological knowledge" (1990, p. 407).

accompany the DNA-sequence information that was originally targeted.[11] On average, any two human beings differ from each other in approximately 10 percent of their nucleotides. Critics argue that any complete understanding of the functions of human DNA must be able to describe and account for this variation (Vicedo 1991).

Consider the meanings of "normal and abnormal" in the context of population genetics theory. Since the state of a population at a given time is given in terms of the distribution of different types of genes, a great deal of information is needed to delineate what is normal and abnormal, including (1) the range of the types of genes, (2) the range of phenotypes and functions associated with these genes, (3) and the range of environments and the related norms of reaction. Finally, some decision about what will count as adequate functioning in a specific environment is also needed; only then can a specific type be categorized as *diseased*. Clearly, such information is not going to be provided by the biochemical causal models that are prominent in molecular genetics.

DEVELOPMENT AND EMBRYOLOGY

A fourth type of biological theory is needed to understand genetically based disease. The models of embryology, epigenesis, and developmental biology, although often confused with biochemical causal-chain models, are distinct from them, because they are designed to describe different things and to answer different questions. Specifically, epigenetic models describe the process of what actually happens with genes in environments; ideally, the end result is a description of the emergence of a phenotype in an environment. In this context, "normal" usually means that you get, at the end of the process of development, what is expected given those genes in that environment. Something is labeled as "abnormal" if it is not what is expected.[12]

The fundamental importance of environmental considerations in interpreting traits as normal or abnormal can be seen in the case of New Guinea highlanders, who often have urinary potassium/sodium ratios four hundred to one thousand times the "normal" Western ratio. Daniel Carleton

[11] E.g., Council for Responsible Genetics (1990).

[12] The distinction between "congenital" and "genetic" birth disorders has long been recognized: "congenital" refers only to disturbances from the normal pattern of development that are not believed to be genetically caused.

Gajdusek argues that this difference is a metabolic response to a sodium-scarce, water-poor environment (Gajdusek, 1970). Notions of proper physiological function, then, depend fundamentally on the related environment.[13]

GENETICS AND DISEASE

Having considered these four types of biological theories, with their corresponding notions of normal and abnormal, we are ready to return to the specific issues surrounding the Human Genome Project. In the genome project, certain genes are labeled as abnormal, and the decision to do so is made by using as a comparison the DNA sequence of a gene that appears in an accepted model of the biochemical causal chain. What is abnormal under the biochemical model is not necessarily abnormal under a medical model. Nonetheless, researchers interested in the genome project routinely slip from a DNA level of description to a medical usage of "abnormal."

P. A. Baird (1990), for example, promises that the genome project will yield a "new model for disease," in which we will be able to diagnose on the basis of causes, and not simply treat symptoms. Carl Cranor points out that Baird's view exaggerates the role of genetic causes in disease (see Cranor 1994). I would add that Baird is assuming the appropriateness of applying the biochemical causal-chain model to all people carrying the gene; the implication is that the gene will produce disease 100 percent of the time, which, as Cranor emphasizes, is very rarely true for genetic disease.

Victor McKusick, former head of the international Human Genome Organization, also tends to overstate the case for genetic causation: "Mapping has proved that cancer is a somatic cell genetic disease. With the assignment of small cell lung cancer to chromosome 3, we know that a specific gene is as intimately connected to one form of the disease as are cigarettes" (McKusick 1989, p. 212). Showing the genetic basis for one cancer is, of course, not the same as showing that every cancer is best understood as a genetic disease.

A similar problem arises in the study of a gene region linked to liver cancer. This study found a region of the DNA where the gene is especially sensitive to exposures to toxins; the toxins induce mutations in that

[13] See the discussion of Gollin et al. (1989).

spot, which then prevent the gene from performing its usual physiological role (Angier 1991). This is a significant advance, especially because this gene has been implicated in many types of human cancer, including tumors of the breast, brain, bladder, and colon. This case seems to support Friedman's claim that "human cancer should be considered a genetic disease" because "it is likely that most human cancer is caused by, or is associated with, aberrant gene expression" (1989, p. 1279). But the toxins appear to be a necessary condition here, in addition to the presence of the sensitive gene region. So, in one sense, the disease is genetic; in another, it arises from environmental causes.

The differences in biological models I outlined earlier can help with this case. Under a biochemical causal model, the mutant gene is a necessary link in the biochemical causal chain of this liver cancer. Hence, the liver cancer has a genetic cause.[14] Under a physiological model, the usual presence of particular proteins is interrupted through the exposure to toxins in the cellular environment. Hence, the abnormal functioning of the body is dependent on the interaction of environment and cells.

The implication in discussions of genetic bases of diseases is that an abnormal gene leads to abnormal functioning, which is itself deficient. But this, of course, is not shown from the strict biochemical description. Both a developmental and functional model are necessary to support the identification of the genetic difference with what we traditionally identify as disease. One problem is that entities on the medical level, such as alcoholism, may be very difficult to pin down genetically. Although a gene believed to be implicated in some severe cases of alcoholism has been found, researchers are more convinced than ever that (1) many genes are involved in the disease and (2) they are *different genes* in different groups of individuals (Holden 1991). As Vicedo has argued, "all the meaningful questions will *start* when all the sequencing is done" (1991, p. 19).

What about the presumption of genetic bases? Many researchers will cite Huntington's disease, or cystic fibrosis, or Down syndrome, as clear-cut cases of disease versus normality. But there are more than a hundred mutations cataloged that will produce cystic fibrosis as a clinical entity; and Down syndrome is diagnosed as trisomy 21, although such a chromosomal arrangement can yield people with a very wide range of abilities. Although some researchers are quick to cite the clear, deterministic cases, out of the total number of genetic screening tests available now, nearly all are

[14] This supports Cranor's (1994) point against Hubbard (1990) that in cases where disease occurs, a single gene can be picked out as a cause.

for gray areas, where the genetic difference is a risk factor, or provides a vulnerability, to develop a specific physiological disease. How is this vulnerability to be understood?

Developmental models are crucial. Gene expression depends inextricably on environment, and environmental responses to knowledge of genetic predispositions can guide development away from dangerous outcomes.[15] Suppose that a person learns she has "the gene for arteriosclerosis" and modifies her diet and exercise regime as a result. We cannot say she will develop arteriosclerosis; in fact, having changed her environmental circumstances, she may well have a *reduced* probability of developing arteriosclerosis in comparison to the population at large. The point is that having a gene for something does *not* imply having that phenotype. "Abnormal" genes may or may not yield "abnormal" or "diseased" organisms. Dr. Henry Lynch, director of a cancer genetics program in Nebraska, is worried that genetics programs will emphasize genetics over simply life-style factors that are much more causally influential (see Roan 1991).

Professor Bernard Davis and his colleagues in the Department of Microbiology and Molecular Genetics at Harvard Medical School have been visible critics of the Human Genome Project. They argue that studies of specific physiological and biochemical functions and their abnormalities will be much more useful medically than the sequencing of the human genome (Davis 1990). Furthermore, only through the refinement and application of developmental biology and population genetics studies of gene distributions can *susceptibility* be studied and interpreted scientifically. Public and scientific misperceptions of susceptibility are probably one of the most prominent problems facing those interested in the development of genetic medicine.

There is a tempting and widespread error in reasoning which is exacerbated by the slippage back and forth between distinct biological meanings of "normal." Under a biochemical causal model, let us suppose a person in whom arteriosclerosis is damaging their health and whose phenotype is clearly "abnormal" and "diseased" according to the medical model. The desired biological explanation traces a causal chain from the

[15] A misleading over reliance on biochemical causal models can lead to the complete disappearance of the influence of environment on phenotype. For example, Watson, in arguing for the sequencing of the complete *Caenorhabditis elegans* genome, says: "If we are to integrate and understand all the events that lead, for example, to the differentiation of a nervous system, we have to work from the whole set of genetic instructions" (1990, p. 48).

genes through the expression and development processes to the resulting pathological state. When asked, "How does arteriosclerosis happen?" the answer is given: "There's a gene for this, which, under these environmental circumstances, takes part in such-and-such a causal chain, resulting in buildup on the arterial walls."

So far, so good. The problem arises when we attempt to understand what it means if a person tests positive for that gene. It is tempting to think that this means they either have or will have arteriosclerosis. But this would be to mistake a contributing cause for a sufficient condition.[16] Exposure to a cold virus is a contributing cause for coming down with a cold, but it is not sufficient; the immune system must also fail to control the spread of that virus in the body. Similarly, having a certain gene might contribute to getting arteriosclerosis, but it is not sufficient; the environmental conditions must also be right in order for arteriosclerosis to become a health problem.[17]

So, take persons with "the arteriosclerosis gene." Are they abnormal? If we define the standard biochemical causal model of fat metabolism as "normal," then they are "abnormal," and the cause of that abnormality is genetic. Are they abnormal on the phenotypic level? Are they diseased? Not necessarily, if we are using the medical model. Inferences that slip from a discovery of genetic abnormality to conclusions of medical abnormality or disease are fundamentally mistaken and unjustified. It is not that the two levels are unrelated or irrelevant to each other; it is just that slipping from "abnormal" in one to "abnormal" in another without evidence is not defensible scientifically.

CONCLUSION

Claims that the Human Genome Project will give us "the recipe to construct human beings" or the keys to understanding "human nature" are misleading at best (Zinder 1990; Smith and Hood 1987; cf. Watson 1990; Vicedo 1991). The usefulness of molecular or DNA-level descriptions by themselves is extremely limited. Genes whose descriptions on the DNA

[16] See the discussion in Hubbard (1990), and Cranor (1994). I believe that Hubbard's view can be interpreted as a rejection of the medical appropriateness of the biochemical causal model owing to the fact that it omits environment as a causal factor.

[17] Another problem with this model of genetic disease is that "diseases caused by a malfunction in one gene tend to be rare," according to molecular geneticists at Harvard Medical School (Davis et al. 1990, p. 342).

level differ from an accepted paradigm of the biochemical causal model may or may not be physiologically significant. Regarding the medical uses of the Human Genome Project, then, the only relevant form of variation is determined by a medical or physiological model. It is important to understand that the medical model of health and disease is just as subject to value judgments as it ever was. Its necessity has not diminished, it has simply gained a wider scope for use. Deciding how human beings *ought* to function is still a negotiated social decision. Proponents of the medical uses of the Human Genome Project have ignored the problems arising from the social nature of disease, but these will not disappear. On the contrary, biotechnology has new powers to implement and *enforce* codes of normality. Molecular biology cannot provide an objective and scientific code of health and normality. The scientific ability to make fine discriminations of variation and the technological power to act on them makes it imperative that *variation itself* be the focus of a searching public debate and educational effort.

9

Evolutionary Psychology

The Burdens of Proof

INTRODUCTION

Richard C. Lewontin's interventions against the acceptance of speculative, untested, yet socially influential claims about human evolution – the most politically significant parts of "sociobiology" – stand as one the most important and controversial aspects of his career. The criticisms expressed in his papers on adaptation (Lewontin 1978, 1985), and in the famous paper he coauthored with Stephen Jay Gould (Gould and Lewontin 1979), have spurred methodological self-awareness about claiming adaptation in many quarters. Although Lewontin's papers have attained the status of obligatory citations, this does not mean that their critical conclusions have been fully absorbed. Indeed, there are recent authors who present G. C. Williams's 1966 book – which was, after all, about the high standards that must be enforced in order to claim an evolutionary adaptation – and Gould and Lewontin's 1979 paper – which embodied an insistence on the high standards that must be enforced in order to claim an evolutionary adaptation – as being on opposite sides of the fence with regard to evolutionary adaptation (Cosmides and Tooby 1995, p. 71; Pinker and Bloom 1992, p. 454).

Chief among those who claim to live in a post-Lewontinian age of adaptationism – one in which an enlightened and modest approach to adaptation is practiced, and the strict standards of scientific evidence enthusiastically adhered to – are those practicing what they call "evolutionary psychology," most prominently, Leda Cosmides and John

Lloyd, E. A. (1999). Evolutionary Psychology: The Burdens of Proof. *Biology and Philosophy, 14*, 211–233. (Prepared for a special volume dedicated to Richard Lewontin).

Tooby.[1] Among the most widely touted experimental evidence for this newly-dubbed field of inquiry is Leda Cosmides's dissertation research on a laboratory reasoning task.[2] From this evidence, she claims to show "how evolutionary biology can contribute to the study of human information-processing mechanisms" (1989, p. 263). These experiments are claimed to provide strong support for her evolution-based "Social Contract theory," and to contradict the most promising non-evolutionary theory of her opponents – the "Pragmatic Reasoning Schemas" of Cheng and Holyoak. Thus, evolutionary theory is used to help design experiments "to discover previously unknown psychological mechanisms" (Cosmides, Tooby, and Barkow 1992, p. 10). Moreover, Tooby and Cosmides have claimed complete empirical victory:

> Our evolutionarily derived computational theory of social exchange allowed us to construct experiments capable of detecting, isolating and mapping out previously unknown cognitive procedures. It led us to predict a large number of design features in advance – features that no one was looking for and that most of our colleagues thought were outlandish (Cosmides and Tooby 1989). Experimental tests have confirmed the presence of all the predicted design features that have been tested for so far. (1995, p. 91)

These claims did not go unnoticed. The paper in which Cosmides's experimental results were reported won the AAAS Behavioral Science Research Prize, and was selected by the judges "because of its substantial and surprising increase in understanding of the rules of thought." (Awards, 1989). The collection of essays coedited by Cosmides, Tooby, and Jerome Barkow, *The Adapted Mind*, was reviewed very positively in many quarters. For example:

> A subtle change of emphasis has been immensely productive. Facts are falling into place all over the field, with implications that stretch far beyond psychology? (*Economist* 1993, p. 82)

[1] The birth of "evolutionary psychology" was celebrated in Tooby (1985), and other early developments in the field include Cosmides and Tooby (1989), Symons (1987), Shepard (1987), and Tooby and DeVore (1987). More recently, we find *The Adapted Mind*, coedited by Barkow, Cosmides, and Tooby, and Steven Pinker's *How The Mind Works*.

[2] "Deduction or Darwinian Algorithms? An Explanation of the 'Elusive' Content Effect on the Wason Selection Task," doctoral dissertation, Harvard University. This research was first published in "The Logic of Social Exchange: Has Natural Selection Shaped How Humans Reason? Studies with the Wason Selection Task" (1989).

> Cosmides and Tooby show that humans have evolved selective mechanisms for detecting violations of conditional rules when these rules mean cheating on a social contract. (Blonder 1993, p. 778)

> Cosmides and Tooby (1987) outlined a theoretically sophisticated and empirically productive method of analyzing human mental capacities as complex biological adaptations, sculpted over evolutionary time through natural and sexual selection. (Miller and Todd 1994, pp. 83–95)

I shall discuss several problems with this research. First, the connections of Cosmides's "Social Contract Theory" to evolutionary biology – contrary to the claims of its promoters – are quite problematic. In addition, it seems that the ostensible links to evolutionary biology – rather than the experimental evidence – are doing much of the work of eliminating rival psychological hypotheses. Once the exaggerated and ill-reasoned claims are removed, the experiments appear to support a non-evolutionary psychological theory at least as strongly: in fact, none of the usual burdensome evidential requirements for an evolutionary hypothesis are even attempted. Evolutionary psychologists are primarily using evolutionary theory to attempt to eliminate other competing theories within psychology, without regard to – and, in fact, in violation of – the standards of evolutionary biology. In what follows, I shall describe briefly the Cosmides theory and its evidence and problems, review updated evidence within psychology, and finally, discuss the relation of evolutionary psychology to evolutionary biology. Part I of this paper will focus on the two leading psychological theories; in Part II, I consider the evolutionary aspects of Social Contract theory.

I should clarify immediately – I am not at all opposed to the application of evolutionary biology to human and animal reasoning, or to psychology more generally. Cosmides's theory and experiments are, in many ways, heading in an exciting direction; to the extent that cognitive psychology has focused on the rules of logic as the ideal form of reasoning, other, more pragmatic, social, or biological bases for reasoning have been neglected. In addition, the general move toward reuniting psychological research and explanation with evolutionary biology is undoubtedly a positive step. My criticisms in this paper reflect concerns that the claims on behalf of the evidence have been overstated, and that evolutionary standards of evidence have been neglected. It is because I think that evolutionary approaches are among the most scientifically promising and robust, that I am concerned about the widespread and less-than-critical acceptance of this body of work. It is one of the most pernicious aspects of the present climate of discussion, that the situation is often set up as a forced choice between accepting the particular theories and oversimplified principles of evolutionary psychology, or retreating to a pre-Darwinian denial of the fact that we are evolved animals.

1. TWO PSYCHOLOGICAL THEORIES

1.1. Cosmides, the Wason Selection Task, and Its Peculiar "Content Effects"

Cosmides's experiments were laboratory tests of the subjects' aptitude at a certain kind of logical problem. This *Wason selection task* is a test of the application of conditional logical reasoning. Experiment subjects are given a conditional rule, with the form, *if p then q*, and are then asked to select which of four given cards must be turned over to decide whether or not the rule holds. Each card has information regarding whether p holds on one side, and whether q holds on the other. So, the subject is presented with four cards facing up: p; ~p; q; ~q, each of which says on the back whether the other variable holds or not.

The correct response according to standard formal logic is to examine the p and the ~q cases (as these are the only ones that could make the conditional false). When the test is put purely in terms of p's and q's, subjects perform poorly, usually neglecting to turn over the ~q card. If, however the rule is, "If you are to drink alcohol, then you must be over eighteen," subjects are significantly more successful in producing the logically correct answer (Cheng and Holyoak 1989, p. 286). Because this difference seems to depend on the content of the rule and not on its form, the resulting difference in performance is called a "content effect." Cognitive psychologists are thus faced with the challenge of explaining why there are content effects, that is, how nonformal aspects of reasoning produce difference in performance on these tasks.

Research into these content effects has been pursued since the late 1960s and several theories have been proposed to explain them. Cosmides presented her social contract (SC) theory as superior to the two leading alternatives: availability-type theories, and the "Pragmatic Reasoning Schemas" (PRS) theory proposed by Patricia Cheng and Keith Holyoak. In fact, she claims that her experiments have the structure of crucial experiments – that is, they don't simply provide positive support *for* her social contact theory, they simultaneously provide evidence *against* the availability and pragmatic reasoning theories.

One of the points of contention about content effects is whether they result from what psychologists call "domain-specific" or "domain-general" cognitive mechanisms. Briefly, domain-general mechanisms are understood as broad mental capacities, which may be used to solve a wide range of cognitive problems, whereas domain-specific mechanisms are much more limited in their ability. Tooby and Cosmides characterize the contrast as follows:

> Does the mind consist of a few, general-purpose mechanisms, like operant conditioning, social learning, and trial-and-error induction, or does it also include a large number of specialized mechanisms, such as a language acquisition device . . . mate preference mechanisms . . . sexual jealousy mechanisms, mother-infant emotion communication signals . . . social contract algorithms, and so on? (1992, p. 39)[3]

Availability theories of the content effect assume a completely general reasoning schema, and they predict that performance on the selection task will depend on familiarity or past experience with the situation. I think it's fair to say that much of the experimental data disconfirms this type of theory; although availability theories claim that performance rests on specific experience, such experience was not found to produce a content effect (Cheng and Holyoak 1985). This leaves us with the two content-specific reasoning theories, Cosmides's Social Contract theory, and Cheng and Holyoak's Pragmatic Reasoning Schemas theory.

On Cosmides and Tooby's account, the ability to perform reasoning about social contracts is evolutionary in origin. According to Cosmides, "a social contract relates perceived benefits to perceived costs, expressing an exchange in which an individual is required to pay a cost (or meet a requirement) to an individual (or group) in order to be eligible to receive a benefit from that individual (or group)" (1989, p. 197). As Gigerenzer and Hug describe the evolutionary rationale:

> For hunter-gathers, social contracts, that is, cooperation between two or more people for mutual benefit, were necessary for survival. But cooperation (reciprocal altruism) cannot evolve in the first place unless one can detect cheaters (Trivers 1971). Consequently, a set of reasoning procedures

[3] Note that on this description, domain-specific mechanisms are considered additions to – and not replacements for – domain-general ones.

> that allow one to detect cheaters efficiently – a cheat-detector algorithm – would have been selected for. Such a 'Darwinian algorithm' would draw attention to any person who has accepted the benefit (did he pay the cost?) and to any person who has not paid the cost (did he accept the benefit?). Because these reasoning procedures, which were adaptations to the hunter-gatherer mode of life, are still with us, they should affect present-day reasoning performance. (Gigerenzer and Hug 1992, p. 130)

Pragmatic reasoning schemas provide a third approach to explaining content effects. According to this theory, pragmatic reasoning schemas – patterns of reasoning – are induced through experience within goal-defined domains (Cheng and Holyoak 1983, 1984, 1985; Cheng, Holyoak, Nisbett, and Oliver 1986; Nisbett, Fong, Lehman, and Chang 1987). Although the schemas themselves are content-dependent, they are created by inductive cognitive processes that are content-independent. The developers of the theory, Patricia Cheng and Keith Holyoak, proposed that

> People often reason using neither syntactic, context-free rules of inference, nor memory of specific experiences. Rather, they reason using abstract knowledge structures induced from ordinary life experiences, such as "permissions," "obligations," and "causation." Such knowledge structures are termed *pragmatic reasoning schemas*. A pragmatic reasoning schema consists of a set of generalized, context-sensitive rules which, unlike purely syntactic rules, are defined in terms of classes of goals (such as taking desirable actions or making predictions about possible future events) and relationships to these goals (such as cause and effect or precondition and allowable action). (Cheng and Holyoak 1985, p. 395)

Cheng and Holyoak investigated a range of content effects in the Wason selection task; their analysis focuses on actions to be taken and conditions to be satisfied:

> If you take action A, then you must first satisfy precondition P
>
> To be permitted to do A, you must first do P

Variants of these rules constitute "permission" and "obligation" schemas, according to PRS theory; these are general schemas set up by experience which guide inferences within specific goal-defined domains. For example, the rule, "If you are to drink alcohol, then you must be over eighteen," is a conditional permission rule. Note that this rule can be mapped onto the material conditional, if p then q. Nisbett and Cheng (1988) discussed combination obligation/permission rules that do not map onto the material conditional, and predicted that performance would follow the pragmatic rule rather than obeying the laws of formal logic.

It was within this tradition of Wason selection task experiments that Cosmides conducted her experiments. She then claimed that her theory was sharply different from Cheng and Holyoak's.

1.2. "Crucial Experiments" and the Elimination of PRS

Cosmides and Tooby have repeatedly attempted to frame the Cosmides experiments as "crucial experiments"; a crucial experiment is one that will, once the results are in, favor decisively one hypothesis over the other alternatives. The goal, with an experimental setup involving a crucial experiment, is to eliminate one hypothesis completely. The experimental results presented in Cosmides's paper should not be understood as the results of crucial experiments. In fact, the experiments do not resolve the competition between Cosmides's Social Contract theory and Cheng and Holyoak's Pragmatic Reasoning Schemas.

Although the experiments comparing availability theory and Cosmides's Social Contract theory can reasonably be interpreted to favor the latter, those designed to compare Pragmatic Reasoning Schemas and Social Contract theory simply do not favor Cosmides's view, contrary to her claims. Rather, evolutionary theory is brought in to tip the balance in favor of Social Contract theory. Given that evolutionary theory thus plays the central role in eliminating competing hypotheses, according to Cosmides's own argument, it is highly significant that her presentations of evolutionary theory are seriously flawed. We will focus briefly on the points at which Cosmides's argument rests on evolutionary theory, emphasizing the centrality of evolutionary theory to her refutation of the clearest and most plausible competing hypothesis.

Cosmides claims that Pragmatic Reasoning Schemas theory has "many theoretical and empirical problems" (Cosmides 1989, p. 193). She begins her attack by reinterpreting their experiments, in which they got a content effect, as social contract problems in disguise. Cheng and Holyoak explicitly gave a social purpose to their tasks, but Cosmides claims that they did more than this – they also gave contextual information, which included a cost-benefit structure. In these cases, then, the permission schemas described by Cheng and Holyoak and Cosmides's Social Contract theory give the same predictions. But Cosmides emphasizes two points of difference between the two theories: the *structure of the proposed algorithms*, and their *origin*.

STRUCTURE OF PROPOSED ALGORITHM In contrast to the problems with testing the competing hypotheses about the origin of the schemas, discussed later, the differences in the structures of the proposed algorithms are supposed to be *directly* testable. These testable differences are described as differences in the "proposed level of representation and domain of operation" (Cosmides 1989, p. 235). In other words, Cosmides claims that her "cost-benefit" analysis involves a different level of abstraction than Cheng and Holyoak's permission/obligation schemas. In Social Contract theory, the benefits and costs are understood in terms of the utilities for the actors. (In evolutionary terms, the utilities must be understood in terms of fitness parameters.)

In PRS theory, in contrast, there are actions to be taken and conditions to be satisfied. But what is the real difference between costs and actions to be taken? Cosmides modifies the meaning of "social exchange" to include cases in which the "cost" is simply "meeting a requirement"; such an expansion of the idea of "cost" violates the notion of exchange, argue Cheng and Holyoak, since a requirement "is not generally an exchangeable entity that can be given in payment" to some individual or group, for the receipt for a benefit (Cheng and Holyoak 1989, p. 288). If we acknowledge that "meeting a requirement" is meaningfully similar to "fulfilling a precondition," then the remaining difference between the two theories is that permission/obligation schemas focus on an "action taken," whereas social contract theory requires that there be a "benefit."

Cosmides argues that a permission rule is a social contact rule only when the "action taken" is a "benefit" and the "precondition" is a "cost." But why call this a different "level of representation"? The issue is simply that the permission rules, according to her analysis, have a larger domain than the social contract rules. Nevertheless, Cosmides proposes critical tests "to decide which kind of representation is psychologically real: the action-precondition representation, or the benefit-cost representation" (1989, p. 237). But this is too strong; these are not the only two choices. It is quite plausible that neither hypothesis under consideration represents something "psychologically real." Later, her conclusions are even stronger. She writes, "cost-benefit representations of Social Contract theory have psychological reality," and she claims to have established that "Social Contract theory posited the correct domain of operation" (1989, p. 253). It is central to her claim that *only* those permission rules having the cost-benefit structure of a social contract actually *work* (1989, p. 254).

Cosmides's strategy seems to be to reinterpret every case in which Cheng and Holyoak found a robust content effect, as actually a social contract in disguise. Cosmides even claims that a content-free permission rule is similarly a social contract.

Cheng and Holyoak's rule is:

> If you take action A, then you must first satisfy precondition P
> To be permitted to do A, you must first do P

Cosmides argues that although this rule does not mention costs and benefits, it still "has an implicit cost-benefit structure." This is because saying that one must satisfy a precondition "is just another way of saying that one must pay a cost or meet a requirement" (1989, p. 239).

But "paying a cost" and "meeting a requirement" or precondition are simply not the same thing. The precondition may be something completely innocuous, with no element of perceived cost to it, or even with an element of pleasure.

Consider the following example: You are permitted to donate blood, if you are HIV-negative. Is the condition of being HIV-negative a cost? No, it is a precondition. This is a genuine permission rule, and not a social contract. There is no sensible way that we would want to interpret this as a cost-benefit relation. Note that it is important to the original logic motivating the "cheat-detector algorithm" that the cost *actually be* a cost; otherwise, there would be no motivation at all to look for cheaters.

ORIGINS OF PROPOSED ALGORITHMS For the sake of argument, let us grant that Cosmides's experiments do produce results in which the participants seem to be using a "cost"-benefit structure in their reasoning. Even so, this result does not eliminate the primary competing hypothesis, Pragmatic Reasoning Schemas theory, which can be expanded to account for these data. (Just as the original Social Contract theory definition of "cost" was modified to include "requirement.") But Cosmides claims that there is a much more basic difference between the two theories. Cosmides ultimately appeals to the greater *a priori* plausibility of Social Contract theory, which is supposedly based on evolutionary theory, to eliminate the competing hypothesis. A crucial part of her argument is that the pragmatic reasoning schemas *cannot explain the origin* of the reasoning mechanism that is used.

Under Pragmatic Reasoning Schemas theory, the schemas come from some form of induction. The rules of inference are a product of experience, which is structured by innate information-processing mechanisms that are domain-general. Under Social Contract theory, in contrast, the rules of inference "are themselves innate, or else the product of 'experience' structured by innate algorithms that are domain-specific" (Cosmides 1989, p. 235).

Cosmides asserts that Cheng and Holyoak's claim regarding the *origin* of schemas is unfalsifiable (1989, p. 235). This is clearly wrong. One could determine experimentally what sorts of experience are relevant for a particular schema, then look for cross-cultural variation in that sort of experience, and finally, determine whether that variation correlates with variation in schema present cross-culturally. One could then test whether inference rules are different.

Furthermore, Cosmides admits that, in her case, the origins claim is not directly testable, but argues that the claim is "subject to plausibility arguments based on existing data." Cosmides claims explicitly that alternative explanations fail because their basic assumption of domain-general mechanisms is false, and she claims that *this falsity is demonstrated by evolutionary theory*. Natural selection, she claims, would have produced special-purpose, domain-specific mental algorithms (Cosmides 1989, p. 193).

The basic structure of her argument is:

1. There are domain-specific processes;
2. These processes are based in evolution;
   Therefore,
3. They are genetically based.

Hence, she is making a claim about the origin of specialized processes, although she admits she's not testing this claim. The evidence she offers for this evolutionary conclusion, in the context of these specific experiments, is: first, that reasoning performance is found to depend on content; and second, that the alteration in performance occurs in the "predicted adaptive direction." She must, therefore, establish *indirectly* that this specificity is genetic, as opposed to being the result of the particular environment or experience; otherwise, there is no real difference between her hypothesis and that of Cheng and Holyoak.[4]

Part II of her paper is meant to address this very question. She must show that the evidence that she has for the existence of specific processes is not a result of experience interacting with a more general social-learning mechanism. Evolutionary theory is used to predict the existence of specific processes. On my analysis of her argument, then, the *gap between her*

[4] Note that Cosmides assumes that genetic and environmental specificity are opposed. This assumption has been criticized extensively by Lewontin (1974, 1985, 1990).

conclusion and the empirical evidence she offers is filled in by the putative power of evolutionary prediction.

The concrete evolutionary claims that are called upon to do this job of eliminating the leading alternative hypothesis include a conclusion stated at the beginning of the paper. Primary among the evolutionary assumptions is *modularity*; in defense of this extremely strong physiological or functional assumption, she cites Noam Chomsky – a linguist known for his hostility to Darwinian explanations – and Jerry Fodor – a philosopher whose speculative book, *The Modularity of Mind*, was deeply embedded in a particular research program in cognitive science, and was notably uninformed by evolutionary thought (Cosmides 1989, pp. 190, 193; cf. Cosmides and Tooby 1992, p. 165). But the main reason for expecting to find extremely specialized modules is evolutionary, according to Cosmides:

> The more important the adaptive problem, the more intensely selection should have specialized and improved the performance of the mechanism for solving it (Darwin 1859/1958; Williams 1966). Thus, the realization that the human mind evolved to accomplish adaptive ends indicates that natural selection would have produced special-purpose, domain-specific, mental algorithms – including rules of inference – for solving important and recurrent adaptive problems (such as learning a language; Chomsky 1975, 1980). (Cosmides 1989, p. 193)

She says that it's clear that although some mechanisms in the cognitive architecture "are surely domain-general, these could not have produced fit behavior under Pleistocene era conditions (and therefore could not have been selected for) unless they were embedded in a constellation of specialized, content-dependent mechanisms" (1989, p. 194). One might ask why they could not have produced fit behavior; this claim seems to assume the impossibility of general mechanisms leading to adaptive behavior. In support of this assumption, Cosmides cites a number of papers in which it is assumed that the problems solved by the cognitive mechanisms in question are: (1) fixed; (2) under strong selection pressure; and (3) have only one solution (1989, p. 195; cf. Cosmides 1985; Cosmides and Tooby 1987; Symons 1987; and Tooby 1985). I need not belabor the fact that none of these assumptions are given empirical support in these papers.

In summary, Cosmides uses the claim that evolutionary theory allows for or predicts her favored type of mechanism to discard or eliminate obvious competing hypotheses. Given the invalidity of her evolutionary assumptions, to be discussed in a moment, the fact that they play a central role in her arguments brings her conclusions into doubt. This does not affect, of course, the fact that there are content effects of a certain, interesting type in her experimental results, but it does bring her conclusions about these results into question.

1.3. Updates and Revisions

Cosmides's initial presentation of her theory and experiments raised a great many questions. Among the most pressing was: what is the real difference between her Social Contract cases, and the permission and obligation schemas proposed by Cheng and Holyoak? Gerd Gigerenzer and Klaus Hug performed an elegant and powerful series of Wason selection task experiments to address this very problem (1992). As Gigerenzer and Hug's experiments showed, the social contract cases comprise one category of Cheng and Holyoak's pragmatic reasoning schemas.

Cosmides claimed that there was a sharp distinction between a Social Contract rule and a non–Social Contract permission rule. She also claimed that evolutionary theory predicted that people would be good at the logic task when there was a SC rule, but not if the rule was a non-SC rule. Gigerenzer and Hug rejected Cosmides's claim that if a rule is perceived as a social contract, then it will produce a content effect. They distinguished experimentally between cases of social contracts – which are, indeed, versions of permission or obligation schemas – and those in which the person is "cued into the perspective of a party who can be cheated" (1992, p. 127). Their experiments provided evidence that people are especially good at detecting cheaters to rules, and that their own role or perspective in the rule affected their performance.

Gigerenzer and Hug showed experimentally that the crucial factor producing a content effect is not whether a rule is a social contract rule; what mattered was whether a subject was cued into the perspective of one participating party in a requirement-benefit exchange, and whether the other party had the option of cheating (1992, p. 165). In other words, the fact that a rule is perceived as a social contract is insufficient for producing a content effect, contrary to Cosmides's claim. Moreover, there seem to be only terminological differences between Cosmides's "requirement-benefit" social contracts, and Cheng and Holyoak's permission/obligation schemas. Gigerenzer and Hug summarize:

> It is the pragmatics of who reason from what perspective to what end (e.g., cheating detection) that seems to be sufficient [to account for the content effects]. Although Cosmides' prediction is at odds with this result, and

> although the distinctions between perspectives, and between social contract rules and Darwinian algorithms were not part of Cosmides' experiments, they nonetheless underlie SC theory. (1992, p. 166)

I find this summary difficult to reconcile with Cosmides and Tooby's later claim that "experimental tests *have confirmed the presence of all the predicted design features that have been tested so far*" (1995, p. 91; emphasis added). The chart provided to back up this sweeping claim curiously omits Cosmides's initial predictions tying content effects to instances of social contracts (1995, p. 90).

Thus, Gigerenzer and Hug credit Cosmides for introducing the notions of the "cheating option" and "the cheater-detection algorithm" into research on this reasoning task, and they note that the notion of cheating itself implies that at least two parties with two different perspectives exist. Given Gigerenzer and Hug's experimental results showing the importance of these distinctions, I think it's fair to say that the Social Contract theory has advanced understanding of experimental conditional reasoning. Nevertheless, nothing in either Cosmides's original data or in Gigerenzer and Hug's data even addresses the underlying evolutionary claims. In fact, Gigerenzer and Hug note that the notion of "cheater-detection algorithm" "could also be derived from points of view other than an evolutionary one, such as from the work on children's understanding of deception as a function of their ability for perspective change (e.g., Wimmer and Perrier 1983)" (Gigerenzer and Hug 1992, p. 130).

The most worrisome arguments offered by Cosmides, from my point of view, involved convoluted discussions of why evolutionary theory favored social contracts and disfavored permission and obligation schemas. Although these have since been abandoned in their original form – because Gigerenzer and Hug's results more persuasively favored part of Cosmides's original hypothesis – specious evolutionary reasoning still appears in Cosmides and Tooby's attempts to eliminate rival approaches.

So, the good news is that a pattern of socially defined costs and benefits seems to enable or facilitate reasoning about social situations. Why this should be a surprise is not completely clear; the answer seems to lie in cognitive psychology's prior commitment to holding deductive formal logic as the natural form of human reasoning. Anyone who has taught formal logic knows that the material conditional violates intelligent and rational individuals' sense of fair reasoning.

The bad news is that these results have been tied to a dogmatic and oversold scientific program. Although Gigerenzer and Hug make very clear, both at the beginning and end of their paper, that other, nonevolutionary explanations of the "cheater-detection mechanism" are plausible (and, I would add, are not even addressed by any of the evidence), Cosmides, Tooby, and other promoters of evolutionary psychology claim that the evidence favors their – and only their – evolutionary account.

2. USES OF EVOLUTIONARY BIOLOGY

Cosmides and Tooby advocate that cognitive science be guided by:

1. Theories of adaptive function,
2. Detailed analyses of the tasks each mechanism was designed by evolution to solve,
   and
3. The recognition that these tasks are usually solved by cognitive machinery that is highly functionally specialized (1995, p. 70).

Evolutionary biology seems so simple, elegant, and powerful; once the power of a selection process to produce evolutionary change is appreciated, it is tempting to apply this process to every situation. But, as evolutionary theorists since Darwin have recognized, evolution involves more than the process of natural selection. Other evolutionary processes – involving chance genetic sampling, various kinds of constraints on variation and development, and phylogenetic history – are ever present, and may even be more powerful than natural selection in the production of a given evolutionary outcome of interest. Admittedly, much of the tradition of evolutionary biology involves focusing on only those characteristics which are primarily understood through selection processes – that is, adaptations – traits that have spread through populations because of their beneficial contributions to that organism's way of living. But this does not mean that every trait – or even most traits – are evolutionary adaptations.

The problem of whether a characteristic is or is not an evolutionary adaptation is even worse in cases of behavioral traits involving learning or higher-level cognitive functions. Consider an hypothesis that I will call the "plasticity" view: the ability to learn the details and subtleties of social interaction – call this "social intelligence" – is an evolved capacity, one under fairly strong selection pressure in social species; moreover, there are aspects of social intelligence that display great flexibility of expression, depending upon social upbringing. I think that there is good supporting evidence for this view from studies of primate behavior (e.g., de Waal 1989, 1991b; de Waal and Harcourt 1992; and esp. de Waal and Johanowicz 1993). Although it is clear that *capacities* for learning patterns of reasoning may have evolved under selection, as is suggested in the "plasticity" approach, and are thus good candidates for being adaptations, it is always difficult to disentangle how much of a given pattern of responses is a part of the biological capacity and how much is the result of the interaction of that capacity with the organism's environment during its growth and development. Given these difficulties – well-known especially since Konrad Lorenz and Nico Tinbergen's pioneering experiments on animal behavior – it is *not scientifically acceptable* within evolutionary biology to conclude that, because a given pattern of responses contributes to evolutionary success, then there is some "organ" (or part of the brain) producing such a pattern, that is therefore an adaptation (see Williams 1966). This is because the "organ" or "module" may not actually exist as a biologically real trait, and even if it does, its current function may or may not be the same as the past function(s).

Cosmides and Tooby (1995) show much evidence of not having understood this scientific standard or its importance. For example, they describe Gould and Lewontin's discussion of the difficulties of demonstrating adaptations, as the view "that natural selection is too constrained by other factors to organize organisms very functionally" (1995, p. 71). But, of course, nowhere have Gould and Lewontin denied that organisms are very functionally organized, or denied the utility of functional reasoning in biology; rather, they have emphasized the complexity of this organization, and the evidential and methodological difficulties in empirically testing functional reasoning.

Cosmides and Tooby claim to avoid the difficult problem of using optimal design as a standard for functional and evolutionary performance: "However, when *definable engineering standards of functionality* are applied, adaptations can be shown to be very functionally designed – for solving *adaptive* problems" (1995, p. 71; emphasis added). This immediately raises the question: How do they know what is an adaptive problem?

> The brain can process information because it contains complex neural circuits that are functionally organized. The only component of the evolutionary process that can build complex structures that are functionally organized is natural selection. And the only kind of problems that natural selection can build complexly organized structures for solving are adaptive problems, where *"adaptive" has a very precise, narrow technical meaning* (Dawkins

> 1986; Pinker and Bloom 1990; Tooby and Cosmides 1990a, 1992; Williams 1996). (1995, p. 76; emphasis added)

The italicized claim will come as a surprise to most evolutionists and philosophers of biology; still, perhaps this is a harmless exaggeration. The pertinent question is whether Cosmides and Tooby have understood the theoretical and evidential standards required when investigating the evolution of structures and processes. And here we see confusion. Take the following passage, for example:

> *An organism's phenotypic structure can be thought of as a collection of 'design features'* – micro-machines, such as the functional components of the eye or liver.... Natural selection is a feedback process that "chooses" among alternative designs on the basis of how well they function. *By selecting designs on the basis of how well they solve adaptive problems, this process engineers a tight fit between the function of a device and its structure.* (1995, pp. 72–83; emphasis added).

Note that there are several issues here, including a basic description of the evolution of an adaptation by natural selection, a claim about the degree to which a selection process will result in an adaptation which fulfils engineering standards ("tight fit"), and a research recommendation for approaching phenotypic traits.

Now look at the footnote to the above passage:

> All traits that comprise species-typical designs can be partitioned into adaptations, which are present because they were selected for, byproducts, which are present because they are causally coupled to traits that were selected for, and noise, which was injected by the stochastic components of evolution. *Like other machines, only narrowly defined aspects of organisms fit together into function systems: most of the system is incidental to the functional properties.* Unfortunately, some have misrepresented the well-supported claim that selection organizes organisms very functionally as *the obviously false claim that all traits of organisms are functional – something no sensible evolutionary biologist would ever maintain.* Nevertheless, cognitive scientists need to recognize that while not everything in the designs of organisms is the product of selection, all complex functional organization is (Dawkins 1986; Pinker and Bloom 1990; Tooby and Cosmides 1990a, 1990b, 1992; Williams 1966, 1985). (1995, p. 73; emphasis added)

Here we see Cosmides and Tooby's ritual recitation of the objections to approaching research in the way they just recommended: *"An organism's phenotypic structure can be thought of as a collection of 'design features'."* I call it "ritual recitation" because there is, in this discussion as well as in other recent writings (e.g., Pinker, Dennett), a peculiar disconnect between what the authors explicitly acknowledge as serious theoretical and evidential problems, and how they actually theorize and evaluate evidence. Let us digress for a moment to consider this problem.

Gould and Lewontin wrote:

> [The adaptationist programme] regards natural selection as so powerful and the constraints upon it so few that direct production of adaptation through its operation becomes the primary cause of nearly all organic form, function, and behaviour. Constraints upon the pervasive power of natural selection are recognized of course (phyletic inertia primarily among them, although immediate architectural constraints . . . are rarely acknowledged). But they are usually dismissed as unimportant or else, and more frustratingly, simply acknowledged and then not taken to heart and invoked. (1979, pp. 584–585)

That is, the problem is not that these other forces of evolution are not acknowledged, it's that they are not taken seriously.

In the discussions that ensued, Ernst Mayr made a good case for treating hypotheses of function and adaptation as reasonable and important investigative tools (1983). Mayr seems right, and his view is also compatible with Gould and Lewontin's position in the "Spandrels" paper. That is, they were not attacking the utility and importance of adaptive hypotheses in research overall; they were attacking the detailed, potentially distorting manner in which these hypotheses were pursued to the exclusion of other potential explanations. As we have seen, this argument proved too subtle for many of its targets to grasp. The issue was always a matter of the *actual weight given in practice* – not in lip-service – to the variety of possible causes of phenotypic traits.

Here is the standard used by Cosmides and Tooby to evaluate whether a trait is an adaptation:

> To show that an aspect of the phenotype is an adaptation to perform a particular function, one must show that it is particularly well designed for performing that function, and that it cannot be better explained as a by-product of some other adaptation or physical law. (1995, p. 90)

But their experiments were not designed to answer evolutionary questions at all: they did not examine whether the "aspect of the phenotype" was, in fact, a well-defined biological trait; they did not examine whether variants of the phenotype were correlated with variants of fitness; and they did not demonstrate whether that aspect of the phenotype was better explained as an adaptation or otherwise. This is why Tooby and

Cosmides appear confused: while they give lip-service to a variety of possible causes of phenotypic traits, only adaptation by natural selection is given actual weight in their practice. And this is done while they claim *not* to be committing the errors in reasoning criticized by Lewontin and by Gould.

But surely the most important issue for other scientists and critical readers is whether Cosmides and Tooby's claims about evolution and their social contract theory are true or not. Here I have some basic worries and questions.

2.1. How Can an Evolutionary Approach Be Based on "Perceived" Cost and "Perceived" Benefit?

> Social contract theory is directly derived from what is known about the evolutionary biology of cooperation, and is tightly constrained as a result. It explains why it should be present in the human mind, what its domain of operation will be, what kinds of implicit inferences it will generate, and what the structure of the "look for cheaters" procedure will be. (Cosmides 1989, p. 233)

Cosmides and Tooby use "constraints" to develop a computational theory of social exchange (Cosmides and Tooby 1989). They claim that any algorithm capable of solving the adaptive problem of social exchange must have certain design features. Among these features is that the algorithms produce and operate on cost/benefit representations of exchange interactions (they cite Axelrod 1984: Axelrod and Hamilton 1981; Trivers 1971). Hence, they conclude, we need, for survival, a cognitive mechanism to assess the costs and benefits of different actions. The algorithm operates on this information, in order to calculate whether the benefits outweigh the costs. Furthermore, this process should be item-independent; it should operate only on the level of the cost-benefit representations. Therefore, the algorithm should be able to handle a variety of items, so long as they are perceived as costs and benefits.

Note that the social contract suggested by Cosmides relates *perceived* benefits to *perceived* costs (1989, p. 197). She argues that providing a benefit doesn't necessarily have a cost, and the benefit doesn't have to actually be a benefit, just "something that he or she considers to be a benefit" (1989, pp. 235–236).

Cosmides can't really mean this. In order for these social interactions to play the role she assigns them in evolutionary processes, the benefit needs to be a *real* benefit, in terms of evolutionary fitness. In other words, the benefit must actually contribute to the probability of the reproductive success of its owner. This is perhaps clearer on the negative side. If the cost were not a real cost, that is, a real cost to the fitness of the genotype, there would be no selection pressure against cheating. In order for Cosmides to tell a selection story about these traits involving cost and benefit, whatever is *perceived* socially as cost or benefit must be positively correlated with fitness. Otherwise, there is no connection at all with evolution. A selection regimen operating on *perceived* costs and *perceived* benefits is not going to produce evolutionary change unless these perceptions are systematically related to *real* evolutionary costs and benefits. It is quite striking that – in spite of the widespread discussion of these results – this has not been pointed out, nor have Cosmides and Tooby made any effort to provide the needed evidence to transform their account into a genuinely evolutionary one.

2.2. Why Is the Standard of "Explicit Instruction" Used When Eliminating Learning Models?

> Social exchange is a universal, species-typical trait with a long evolutionary history. We have strong and cross-culturally reliable intuitions about how this form of cooperation should be conducted, which *arise in the absence of any explicit instruction* (Cosmides and Tooby 1992; Fiske 1991). (Cosmides and Tooby 1995, p. 77; emphasis added)

This absence of explicit instruction is then used as evidence that this pattern of behaviors or preferences is not learned, and hence is not a product of a more general social-learning ability, such as the "plasticity" view discussed in the previous section. But surely the presence of "explicit instruction" is not a legitimate prerequisite for a task being learned; the development of highly complex social skills in non-human primates is not associated with much explicit instruction at all (see de Waal 1989, 1991a; de Waal and Johanowicz 1993; Wrangham et al. 1994). Cosmides and Tooby claim that *learning* is not needed, but they have not supplied or evaluated evidence for and against this claim. As Gigerenzer and Hug clearly recognize, all of these experiments are insufficient for eliminating other, nonevolutionary explanations, such as Pragmatic Reasoning Schemas, or some other general social-learning theory. Cosmides and Tooby seem overeager to eliminate rival psychological hypotheses that are based on more multipurpose "modules" than the "highly functionally specialized" ones that they theorize (1995, p. 70). Although they claim that evolutionary biology favors their more specialized modules, there is nothing either in their discussion or in evolutionary theory that supports this assertion.

2.3. Why the Pressure on Efficiency? Why Not Just Look at All the Cards?

In the Wason selection task, the subject is asked to turn over all and only the cards necessary to find out if the conditional is false; turning over *all* the cards is implicitly discouraged by the directions of the experiment. When we consider the test as reflecting an important type of evolutionary scenario, though, why assume that such efficiency of thought was selectively desirable? Why not just turn over all the cards? Why not get as much information as you can, and then decide? Of course, it is not impossible that speed of deliberation might have been a factor in our evolutionary past, but the burden of proof is on those who claim that it was; there is no justification offered for it being assumed. It seems to me likely that the efficiency requirement is simply a remnant from the original context of the Wason selection task – as a test of logical competence, within a nonevolutionary approach in psychology.

2.4. Why Assume that Social Intelligence about Social Exchange Evolved during the Pleistocene?

The evolutionary psychology research program, as outlined by Cosmides, Tooby, Barkow, and Symons, rests on taking the Pleistocene environment as the crucial evolutionary environment. The thought is that the hominid lineage evolved during this period into the basic modern human being; thus, features of our minds that are distinctly human must have evolved during this transition from prehuman to human. But is it reasonable to think that practices of social exchange are distinctly human? Evidence of extremely sophisticated social intelligence has come from our closest relatives, the chimpanzees and bonobos, some of which seems to suggest paradigmatic social exchange behavior (de Waal 1991a, 1991b; de Waal and Harcout 1992; de Waal and Luttrell 1988; Wrangham et al. 1994). Wouldn't the chimps and bonobos also need to have specialized modules, for the same reasons that human beings did? One reviewer of *The Adapted Mind* proposes that the social exchange modules evolved in the common ancestors of chimps and human beings, about six million years ago, thus making "the environmental conditions and lifestyles of hunter-gatherers in the Pleistocene . . . [of] . . . limited evolutionary significance" (Mithin 1997, p. 102).

3. CONCLUSIONS

I applaud the fact that academic psychology is finally being reunited with evolutionary biology; the authors criticized in this paper deserve credit for bringing overdue consideration of evolutionary issues to the fore. Much credit also goes to Cosmides for her pursuit of a particular focus on cost-benefit reasoning; the recent evidence confirms at least some other initial ideas, that is, that some human beings are especially good at such reasoning, and that it facilitates performance on certain logical tasks. Perhaps the most disturbing aspect of this research program is that its originators and defenders overreach. If the evidence is so good, why weaken your position by mischaracterizing it as a crucial experiment to eliminate rival theories? If you're not presenting any empirical evidence that would support a claim of evolutionary adaptation, why say that you are? Why make particular evolutionary claims at all, when they are not tested?

I draw several morals from the story.

First, the lack of basic consensus in psychology has made the field ripe for the sort of exaggerated claims made by these evolutionary psychologists. Second, a lack of awareness of the real standards of evidence in evolutionary biology can lead to acceptance of unwarranted or highly controversial claims, among psychologists seeking a definitive way of choosing among the competing hypotheses offered for the content effects of the Wason selection task. Philosophers of mind suffer the same risk.

Again, it is essential to reject the false dichotomy that is frequently set up rhetorically in these discussions: the choice is *not* between accepting the particular evolutionary psychological proposal under consideration, or rejecting an evolutionary approach to psychological or social traits altogether. These researchers have taken a step in the right direction; this doesn't mean we must follow them all the way down the path.

Finally, this case provides an opportunity to reflect on just how important it is that scientists in other fields be vigilant about the points insisted on by Lewontin and other evolutionary biologists. Virtually all of the writers in evolutionary psychology dutifully cite Lewontin's work, and state that, of course, they are not doing what he warned against. Our apparent human affinity for design explanations and stories about our species has many dangers, chief among them that we don't evaluate adaptive evolutionary explanations as carefully as we should. This was Lewontin's warning in the 1970s against the excesses of sociobiology; evolutionary psychology is in peril of suffering the same scientific ignominy unless its promoters clean up their act.

ACKNOWLEDGMENTS

I would like to thank Carl Anderson, Mark Borrello, Greg Ray, David Smith, and Rasmus Winther for research assistance during various segments of my investigation of this work. Marc Feldman, Greg Ray, and Dan Sperber contributed substantively to this discussion. As this critical project was in progress for nine years, there are many other people to be thanked, including audiences at the CNRS in Paris, Brussels, and UC Davis, as well as my seminar participants at Harvard in Spring of 1998. Finally, so much is due to Dick Lewontin that I refuse to embarrass him by detailing the debt.

10

Objectivity and the Double Standard for Feminist Epistemologies

On the face of it, feminism, as a political movement or ideology, is irrelevant to *truth*. Therefore, it is irrelevant to *objectivity*, which is about truth and how to get at it.

I believe that philosophical and scientific views regarding "objectivity" are the source of the fiercest and most powerful intellectual and rhetorical weapons deployed against feminist critiques of epistemology and of the sciences. This is because philosophy of science and epistemology are, after all, concerned with analyzing cases of *good* reasoning. The philosophical challenges are to formulate and examine: how good scientific knowledge is produced; how and whether other forms of knowledge (e.g., moral knowledge) differ from scientific knowledge; and how to *explain why* science seems to be such a successful way to produce knowledge. The concepts of truth, objectivity, and evidence are at the heart of these investigations, and rightly so, I believe.

Many philosophers acknowledge – under the pressure of overwhelming evidence – that sex and gender issues may play roles in the social sciences, but never in the mathematical or natural sciences.[1] The fact is that

[1] The distinction between sex and gender can be understood for our purposes as follows: *sex* (male/female) refers to biological categories, based on external and internal anatomy, and hormonal and chromosomal combinations: *gender* (masculine/feminine) refers to variable traits and behaviors associated with male or female persons. Even if we make the assumption that sex is not socially constructed and enforced (contra Fausto-Sterling 1993), we cannot say the same for gender, this shows in the variability of the actual contents of gender roles across cultures, and in the frequent mismatch of sex and gender. Basically, the social construction and enforcement of gender roles amounts to the fact that biologically female bodies are perceived as, and socially constructed to have, a culture's "feminine" traits and social roles, whereas biologically male bodies are similarly constructed to be "masculine."

Lloyd, E. A. (1995). Objectivity and the Double Standard for Feminist Epistemologies. *Synthese, 104*, 351–381.

detailed arguments from *within* the scientific community about the influence of sex and gender issues in the natural and mathematical sciences have been around for more than a decade.[2] Yet this evidence has been largely ignored. Instead, many philosophers assume that there *is no* evidence – and *could not* be such evidence – to support feminist analyses of the importance of sex and gender in every branch of knowledge. The vast majority of philosophers still believe that "feminists" are playing "out-of-bounds," in terms of mainstream understandings of the problems of epistemology and philosophy of science; feminist work can, therefore, be safely ignored, set aside, or characterized as of interest only in marginal cases.[3]

Given the recent outpouring of feminist writings that challenge, revise, and apply particular notions of *objectivity*, it is past time to place the burden of proof for the typical philosophical belief – that feminism is irrelevant to the study of the objectivity of knowledge – where it belongs.[4] Specifically, philosophers with views that acknowledge the importance of various human, social, or contextual elements in meaning, concepts, and knowledge-development, are obliged to justify their a priori exclusion of sex and gender as relevant factors. Contemporary social scientists, although they may agree about virtually nothing else, do agree that among the most crucial distinctions in all human societies and cultures are sex and

[2] E.g., Bleier 1984; Conkey 1991; Fausto-Sterling 1985, 1993; Harding 1986, 1991, 1992, 1993b; Hubbard 1990; Keller 1985; Lewontin 1990; Longino 1979, 1990, 1993a, 1993b; Longino & Doell 1983: Potter 1993; Rouse 1987; Traweek 1988; and Wylie 1988, 1989.

[3] This latter conclusion is freely drawn in the context of professional conversations and seminars, but it appears rarely in print (but see Searle 1993). Searle claims that social reformers such as feminists have been "blocked" in analytic philosophy "by a solid and self-confident professorial establishment committed to traditional intellectual values" (1993, p. 71); I argue that Searle and other philosophers have themselves been blocked, through their own "self-confidence," from fulfilling their commitment to these same traditional intellectual values – much to the disappointment of the feminists in question. In addition to anecdotal evidence, though, I would like to emphasize the abundance of public, printed evidence for the existence of philosophical segregation of feminist views, such as citation analysis, lack of response to relevant problems raised by feminist authors, as well as more overt marking and marginalization of "feminist" topics in APA and PSA programs. Make no mistake: I am not making a psychological or motivational claim about my colleagues, but rather, challenging our philosophical communities' acceptance (perhaps unconscious) of documentable, quantifiable social facts.

[4] For feminist discussion of "objectivity," *see* Dupre 1993; Duran 1991; Haraway 1989, 1991; Harding 1986, 1991, 1992, 1993b; Haslanger 1993; Keller 1985; Longino 1979, 1990, 1993a, 1993b; Longino and Doell 1983; L. H. Nelson 1990, 1993; Rouse 1987; Tuana 1989; Wylie 1989. The collections of Alcoff and Potter 1993 and Antony and Witt 1993 are especially useful.

gender.[5] As central organizing principles of all human social groups, sex and gender categories and roles provide the structural underpinning of virtually all *other* social roles, interactions, and complex human activities, such as communication, enforcement of social norms, and standards of behavior. Because sex and gender distinctions serve this foundational role on which the rest of the social structure is dependent, philosophers who wish to take the interests, values, and goals of the relevant knowledge-communities into account – but who nevertheless exhibit an a priori dismissal of the relevance of the central sex and gender distinctions to epistemological and metaphysical problems – must be able to provide, on pain of irrationality, reasons supporting that dismissal.

The problem is that there *is* evidence that sex and gender do indeed play central roles in all forms of human knowledge. These groundbreaking and pivotal arguments have been reviewed elsewhere repeatedly – I will not do so here. This evidence can be resisted only under assumptions which have explicitly been repudiated by a wide variety of contemporary metaphysicians, epistemologists, and philosophers of science. In this paper, I review the positive views of several influential analytic philosophers, and show that they in fact *provide for* the relevance and legitimacy of considerations of sex and gender.

I must assume that this will come as a shock to them, given the invisibility of such analyses in their own work. But such intellectual and philosophical irresponsibility ought not be tolerated or perpetuated. The burden of proof is on these authors – and those who agree with the basic assumptions they hold – to address and argue against both the specific positive claims made by feminist critics, and the general claim that sex and gender are *always* philosophically relevant.

On my analysis, much of the neglect and negative reaction among philosophers to discussions of the roles of sex and gender in knowledge has its source in a specific *philosophical folk story* about *objectivity* and its relation to scientific knowledge which is part of a philosophical tradition.

[5] Conkey 1984, 1991; Leacock 1977, 1978a, b, 1981; Leacock and Nash 1977; Leacock and Safa 1986; Lévi-Strauss 1956, 1969; Reid 1970; Reiter 1975. While Leacock acknowledges that "institutionalized specialization by sex must have been critical somewhere along the line of human emergence," she notes that the specifics of these sex role divisions of labor vary considerably from culture to culture (1981, p. 229; cf. Siskind 1978). See also Frieze et al., who note that "all cultures use sex as one [criterion] for assigning roles," although "in all known cultures," "no matter what specific activities men and women engage in, the roles played by men are valued more by society than the roles played by women" (1978, pp. 79, 83).

One of my aims here is to argue that the anachronistic view of "objectivity" embodied in this folk story has, in general, cast a shadow of confusion over philosophical discussions of reality, knowledge, and language; furthermore, it has been particularly important in obscuring the significance of sex and gender analyses in mainstream epistemology and metaphysics. I shall describe that folk story in a moment, and mention a few crucial problems with it. Then I shall review some of the ways that that philosophical folk story has been resisted, focusing on various twentieth-century philosophers who have actively recast the meanings of "objectivity," through their emphasis on contextual understandings of meaning, truth, and inquiry. But let us start with the basics, and take a look at some of the things that we think of when we consider the terms *objective* and *objectivity*.

1. THE MANY FACES OF OBJECTIVITY

1.1. Four Basic Meanings

I have identified four distinct meanings of "objective" and "objectivity" that are currently in broad use in contemporary philosophy. I focus on these views because they are often mixed and matched into specific hybrid notions of "objectivity" that play central roles in current analytic epistemology, metaphysics, moral philosophy, and philosophy of mind.

Sometimes:

- *Objective* means *detached*, disinterested, unbiased, impersonal, invested in no particular point of view (or *not having* a point of view);
- *Objective* means *public*, publicly available, observable, or accessible (at least in principle);
- *Objective* means *existing independently* or separately from us;
- *Objective* means really existing, Really Real, the way things really are.

Note that when "objective" or "objectivity" play these various roles, they *are predicated of different entities*, depending on their philosophical usage. For example: *detachment* is a property of a knower, and not a property of what is known; *ontological independence*, in contrast, is a *relation* between reality and a knower; *publicity* is also a relation between reality and knowers; and *Really Real* is the status of what *is*, regardless of its relation to any knower.

Consider a few examples that suggest that these four meanings are, indeed, distinct and separable: my consciousness is Really Real, but it's not an immediately accessible public phenomenon, I cannot be detached from it, but it does exist independently for everyone else; the optical illusions of the pools of water on a desert highway are Really Real – they exist in our experience – and they are also public, but they do not exist independently of us as knowers or perceivers; God, if such a being exists, is Really Real, it exists independently of us as knowers, and it is not always presumed to be public; the number three is public, and, in a sense, it is independent of each of us as knowers; still, whether it is Really Real, or whether it is independent of all of us, remains open to debate; finally, plants and flowers are public, they exist independently of us as knowers, and are Really Real.

What *kinds* of philosophical and/or scientific virtues are exemplified under each of the above four meanings?

1.2. Epistemological and Methodological Meanings

"Objective" often involves a particular type of *relation* between the knower and reality-as-independently-existing. The phrase "existing independently from us" can be interpreted in a number of ways; the most common philosophical meaning involves us as human knowers and/or experiencers; often, the knower, and even experiencer, are depicted as minds (as distinct from bodies, in a Cartesian sense, e.g., Nagel 1993, p. 37). Sometimes the emphasis is placed on independence from human will and wishes – that is, on things or events over which we have no individual or human control. In all cases, though, this relation is taken to necessitate some sort of *detachment* of the knower as a methodological virtue;[6] if one is personally invested in a particular belief or attached to a point of view, such inflexibilities could impede the free acquisition of knowledge and correct representation of (independent) reality – otherwise, we cannot distinguish things-as-they-are from things-as-they-appear-to-us as human or as particular observers. Keep in mind that this positive value placed on detachment is derived from a particular epistemological picture, in which "objective knowledge" is defined as public, and as involving existences independent of us.[7]

[6] E.g., Bennett 1965, pp. 6–12; Haskell 1990, p. 133; Johnson 1987, pp. xxiii ff., 200–202; Mackie 1976, pp. 20, 33–34; Nagel 1986, pp. 4, 7, 13–15, 1993, p. 41; Putnam 1981, p. 54; Rescher 1992, p. 188; Searle 1992; Shapere 1981, pp. 38, 47; Wiggins 1976, p. 336; Williams 1985, pp. 1 II, 136, 138–139, 149; and Wright 1992, pp. 1–3.

[7] Authors who endorse various methods of detachment on epistemological grounds include: Haskell 1990, p. 134; Johnson 1987, pp. xxx ff; Longino 1993a, p. 261; Nagel 1986, pp. 4, 5,

"Objective" sometimes means that a phenomenon is *public* or in principle publicly accessible.[8] Publicity of a phenomenon may be interpreted in a weak sense, in which the average-equipped human being will have typical experiences when exposed to a particular stimulus. Publicity of a phenomenon is sometimes distinguished from its accessibility to individuals, through the stronger requirement of "third-person accessibility"; to qualify as *third-person accessible*, it is not enough that there be a relationship between a knower and a phenomenon – another person must be able to experience directly that phenomenon, as well.[9] The point of this requirement is to exclude "first-person" or "private" experiences as potentially *objective*, by definition. For example, it may be true that each of us has particular, personal experiences while swimming in the ocean, but the qualities of that experience, what-it's-like-for-me-to-swim-in-the-ocean, are not third-person accessible; in that sense, the contents of the experience are not public.[10] The philosophical weight put on third-person accessibility is probably associated with the scientific requirements of repeatability of experiments and observations;[11] even if one goes swimming in the ocean, it is difficult to compare directly that what-it's-like with any *other* person's what-it's-like-to-swim-in-the-ocean. Of course, publicity of a phenomenon is insufficient to establish that it exists independently of us – optical illusions are notoriously both public and ontologically dependent – but publicity is often taken to support, under certain conditions, the independent existence of whatever is public. Hence, it is epistemically invaluable.[12]

14, 17, 35–37; Novick 1988; Rescher 1992, p. 190; Scheffler 1967, pp. 1, 8–10; Shapere 1981, p. 45; Williams 1985, pp. 111, 138–140; and Wright 1992, p. 6.

[8] See Dennett 1978a, 1991; Hacking 1983, p. 133; Haskell 1990, p. 133; Johnson 1987, pp. xxvii, 173–174, 190–195; Kuhn 1992, pp. 4, 11; Longino 1993a, p. 264; Quine 1981, pp. 71, 177, 184; Rescher 1992, p. 188; Scheffler 1967, pp. 8, 10; Williams 1978, p. 245, 1985, pp. 139–140; and Wright 1992, pp. 6, 10–11.

[9] Some authors require publicity of the most robust sort for any "real" entity (e.g., Mackie 1976, pp. 18–19; Quine 1981, pp. 2–3, 18–19, 35, 87, 177, 184; Williams 1978, p. 247). Dennett is a particularly fierce advocate of the necessity of third-person accessibility for scientific phenomena (1978a, pp. 151, 154; 1988, pp. 47–49, 1991, pp. 71–74).

[10] Cf. Searle 1992, pp. 94–100.

[11] This public process could result in a *kind of detachment*, in the sense of the neutralization of personal idiosyncrasy, in some cases. Daston 1992 places the concern about neutralizing personal idiosyncrasy in science firmly in the mid-nineteenth century, and associates it with the scientific division of labor and demands on public communication that developed during that period (see, for example, Peirce 1868, 1877; Daston 1992; Daston and Galison 1992; Longino 1993a, b).

[12] E.g. Quine 1974, pp. 36–39; and Williams 1985, p. 139. Historically, see Boyle 1744; Hobbes 1651; and Shapin and Schaffer 1985.

It is vital to distinguish the publicity of the phenomena themselves – whether on an individual or third-personal basis – and the publicity of *standards* that are embodied in a community, that is, publicity of community standards and practices and the shared standards of judgment.[13] Emphasizing the publicity of standards allows personal (i.e., non-third-personal) things to count as "public."

1.3. Ontological Meanings

There are several distinct and relevant ontological meanings of "objective," as well:

"Objective" phenomena are those *existing independently and separately from us as knowers*. "Subjective" phenomena, including "subjectivity" itself, are understood as those phenomena which do *not* – and in some cases, *cannot* – exist independently and separately from us and/or our experience.

"Objective" also means *Really Real*, concerning *the way things really are*.[14] The relevant contrast to *Really Real* is to something that *isn't Really Real*, considered completely independently from our epistemic position about it. We can get a sense of what it is *not* to be Really Real by considering an historical example from the natural sciences. Take the ontological status, *simpliciter*, of spontaneous generation. Although we once assigned this process the status of being *Really Real*, we no longer do so. Of course, it's only our assignment that's changed, but the status of *being Really Real* can be understood independently of any given judgment (*pace* Wittgenstein, Davidson, Putnam and other cryptoverificationists).[15]

[13] See Hacking 1988, pp. 149–151; Johnson 1987, p. 212; Kuhn 1977, 1992; Longino 1990, 1993a, pp. 259–264, 1993b; McDowell 1979, pp. 339, 341; Putnam 1981, 1988, Ch. 7, 1992, pp. 84, 102–103, 214; Quine 1974, pp. 36–39; Rorty 1986, 1988, 1989; Scheffler 1967, pp. 1, 3; Wiggins 1976, pp. 343, 353–360; and Williams 1985, pp. 97–98, 218.

[14] Examples can be found in: Longino 1993a, p. 261; Mackie 1974, pp. 103, 109–112, 223, 285, 1976, pp. 18, 20; McDowell 1979, p. 347, 1988, p. 170; Putnam 1992, pp. 93, 100, 102; Quine 1981, pp. 35, 82, 177; Rescher 1992, p. 189; Rorty 1988, pp. 49, 59–62; Scheffler 1967, p. 10; Williams 1985, pp. 111, 132, 135, 139, 152; and Wright 1992, pp. 1–6. John Searle is an exception; he distinguishes between "ontologically objective" and "ontologically subjective," both of which are real (1992, pp. 94–100).

[15] Those who offer some version of the argument that we cannot really "grasp" or make sense of such a concept include: Davidson 1967, 1974; Hacking 1983, pp. 55, 110–111, 1988, p. 151; Putnam 1981, esp. pp. 51–55, 1992, pp. 101–103, 108; Quine 1981, pp. 21, 97; Rorty 1979, pp. 344–345, 1982, 1986, 1987, pp. 36–45, 1988, pp. 54–56, 62–63, 68, 1989; Scheffler 1967, pp. 34–35; Strawson 1959, p. 128; Wiggins 1976, esp. pp. 350–353, 1980; and

2. THE ONTOLOGICAL TYRANNY

Let us now examine the strong claim that "objective" reality – the reality converged on through the application of objective methods – equals all of the Really Real.[16] Such an equation seems to be a judgment that the Really Real can, in its totality, be reached or known through its being publicly accessible in the proper way, combined with the right sort of detachment of the inquirers. I call this position the *ontological tyranny*.

2.1. The Mutual Support of Methods and Ontology

The *ontological tyranny* – the position that "objective methods," as sketched above, provide our *only* legitimate access to the ontology of the Real – can only work if the Really Real is also *completely* independent of us. And this only makes sense if we have *prior* commitments to a particular sort of ontology. If we seek objective knowledge, and "objective" reality is defined as that which exists completely independently from us, then an effective epistemology will involve the *removal* of any attachments or points of view that might interfere with our *independence from* the reality we wish to know. This gives rise to the epistemological requirements of: (1) the public accessibility of objects of knowledge; and (2) our detachment as knowers.[17] *Publicity*, as an epistemological requirement, involves two ontological facts – that "objective reality" exists independently from us, and is *publicly* knowable, if it is knowable at all (e.g., Mackie 1976, pp. 20–23).

An ontological aspect of *detachment* is also involved in this method. Once objective knowledge of reality is identified as knowledge of independent existences, and knowledge of independent existences requires

Wittgenstein 1953 (under some interpretations). Arguments against such a conclusion are offered by Williams (1981, 1985, pp. 138–140), Stroud (1984), and Nagel (1986, pp. 37, 105–109).

[16] See P. M. Churchland 1989; P. S. Churchland 1986, 1993; Haskell 1990, p. 134; Mackie 1976, pp. 18–20; Nagel 1986, pp. 7, 26, 77; Novick 1988; Quine 1969, 1981; Rescher 1992 p. 189; Smart 1963, pp. 84, 92–95; Stich 1983; Teller 1992. Dennett (1978a, pp. 255–256, 1988, 1991) insists that a "proper" ontology of mind and morals will include only ontologically independent and publicly accessible things (see also Quine 1969, 1981, pp. 10–21, 90; Sellars 1968, pp. 133–134, 150).

[17] See Haskell 1990, p. 134; Johnson 1987, pp. xxx ff.; Longino 1993a, p. 261; Nagel 1986, pp. 4, 5, 14, 17, 35–37; Novick 1988; Rescher 1992, p. 190; Scheffler 1967, pp. 1, 8–10; Shapere 1981, p. 45; Wiggins 1976, p. 340, 343; Williams 1985, pp. 111, 138–140; and Wright 1992, p. 6.

the knower's independence from those existences, *detachment* becomes a methodological virtue. That is, the ontological assumption that reality is that-which-is-independent-from-us, serves as the rationale for a *method* of detachment.[18]

Historically, one ontology that identified the Really Real with complete independence from knowers, consisted in the claim that Primary Qualities are real in matter *(res extensa)* in a way that Secondary Qualities are not. The physical corollary was that *there is nothing but matter in motion*; a modern equivalent might refer to *energy/matter* or *fields*, and so on, but would still deny equal ontological status to properties that emerge only in the context of our interactions with the physical stuff.

Hence, not only does the *ontological tyranny* dictate only one method as adequate, it also identifies only one set of objects as possible objects of objective knowledge.

2.2. *Problems with the Ontological Tyranny*

This particular set of connections among meanings of "objectivity" – in which the meaning of "objective reality" is interpreted as "the way things are independently of us and our beliefs" – serves as the justification and motivation for an "objective method," where this includes: *detachment*, in the sense of removal or abnegation of our internal lives, the "subjective content" of our experiences; and *publicity*, in the sense of the direct public observability of the phenomena in question. This web of interdependent meanings – the ones that tie an epistemological to an ontological meaning, which together justify a method – was itself, as a *cluster*, born and defended – like any other idea – in a specific historical context.

The *ontological tyranny* played a central role in seventeenth- and eighteenth-century philosophy in characterizing the differences between Primary and Secondary Qualities. These philosophical doctrines were born in the pursuit of the physical sciences; they soon, however, took on a life of their own, and were embalmed, especially in the context of eighteenth-century philosophy of mind, into a *standard for knowledge*

[18] We may ask what, exactly, is meant by a method of "detachment," and what its role is in being able to know things independent from us as knowers; there is some question whether a coherent understanding of "complete detachment" is possible. One solution lies in conceiving detachment as having degrees; the relevant claim here is that the greater degree of detachment will give us greater knowledge of things independent of us, other things being equal.

itself, which I call the "philosophical folk view of objective knowledge."[19] This folk story involves particular views about Primary and Secondary Qualities.[20]

Primary Qualities are absolute, constant, and mathematical; they make up the *true* material objects – the *Real* – and the domain of knowledge, both human and Divine. The Primary Qualities – extension, number, figure, magnitude, position, and motion – have two crucial properties: they *cannot* be separated from bodies; and they *can* be represented wholly mathematically.

Secondary Qualities are, in contrast, fluctuating, confused, relative, and untrustworthy; they arise *from* the senses. They are merely the effects on our senses of the Primary Qualities, which are alone the qualities of real nature.[21] Secondary Qualities aren't as real as Primary Qualities, precisely because they depend for their very existence on some sensing being.[22]

Of course, historically, it was not epistemological independence itself that *motivated* the separation into Primary and Secondary Qualities; the motivation was, instead, clearly ontological: The Reality of the physical universe is geometrical.[23] Primary Qualities were promoted ontologically to first place *because* they were unchanging and constant, and therefore, subject to mathematical representation.[24]

19 This afterlife in philosophy arose in spite of the early death of the doctrine of Primary/Secondary Qualities in the physical sciences themselves (Hesse 1974, p. 286; McMullin 1988b, p. 233; Stein 1993). In fact, there was no univocal view, among seventeenth- and eighteenth-century investigators, about Primary and Secondary Qualities; the historical record certainly shows a wide range of views. Compare Boyle 1744; Hobbes 1651, 1839; Locke 1694; Berkeley 1871; and Newton 1726, 1972; or see commentaries by Curley 1972, 1978; Garber 1978, 1992; Grene 1985, 1991; Stein 1993; Stroud 1980; Williams 1978, pp. 237–239; and Wilson 1978, 1979, 1982, 1993.

20 This is, of course, only one of the available interpretations of Primary and Secondary Qualities, but its variants have been the focus of attention in contempory philosophy.

21 For instance, Locke, in arguing for mechanical explanations (which, depending on which parts of Locke you read, include only Primary Qualities, or also Secondary Qualities, when understood as dispositions or powers based in Primary Qualities to evoke certain sensations in observers), said that they provided "the only way which we can conceive bodies to operate in" (1694, II, pp. 8, 11). The emphasis on the Subject so central to both Montaigne's and Descartes' projects – had virtually disappeared, even at this early date.

22 Descartes, for example, required that the objects of physical inquiry be completely independent of our knowledge, will, or experience (see Curley 1978, pp. 147–149; Grene 1985). *See* Locke 1694 II, pp. 8, 24–5; Bennett 1965, pp. 12, 14–17; Curley 1972, p. 442; and Mackie 1976, pp. 11, 16.

23 See Bum 1932; Curley 1978, pp. 4–9; Grene 1993, pp. 76–77, 84; and Williams 1978, p. 236.

24 Most conspicuously, Galileo and Descartes. In the Fifth Meditation, Descartes describes the physical Really Real: "I have ideas of things which, whether or not they exist, and

To make this very clear: part of the view embodied in the web of interlocked meanings of philosophical folk-objectivity originally grew out of particular historical individuals and communities with particular projects involving physics. Most importantly, each of the original historical individuals and communities involved had *prior ontological commitments* that served as the rationale and justification for their adoption of the specific methods deemed appropriate for investigating nature.

They were very explicit about this. Here's Galileo:

> Philosophy is written in that great book which ever lies before our eyes – I mean the universe – but we cannot understand it if we do not first learn the language and grasp the symbols, in which it is written. This book is written in the mathematical language, and the symbols are triangles, circles, and other geometrical figures, without whose help it is impossible to comprehend a single word of it, without which one wanders in vain through a dark labyrinth . . . (Bum 1932, p. 75)

Galileo's commitment to a mathematical science was based, in turn, on a religious view, namely, the idea that God made the world an immutable mathematical system, which permits an absolute certainty of scientific knowledge by human use of the mathematical method: "As to the truth, of which mathematical demonstrations give us the knowledge, it is the same which the Divine Wisdom knoweth . . ."[25]

In other words, Galileo and Descartes, having committed themselves to an ontology that was fundamentally religious, then 'discovered' that Secondary Qualities are dependent on us, and reasoned, from their ontology, that they must therefore not be Really Real in the physical realm. Hence, I agree in part with Colin McGinn's claim that the fact that Primary Qualities are *independent* from us is a *philosophical discovery* – rather than being a *scientific* one – although I am hesitant to call a religious dogma a "discovery" (McGinn 1983).

In summary, the configuration of ontological and methodological commitments that constitute the standards of "objectivity" for early modern physical sciences explicitly relies on religious beliefs.

whether or not I think of them, have *true and immutable natures* or essences" (Curley 1978, p. 141; Descartes VII, p. 84; my emphasis). Compare Quine, who, in endorsing the philosophical currency of the Primary/Secondary Quality distinction, describes "subjectivity" as varying, inconstant, not "true once and for all" (1981, p. 95).

[25] Galileo Galilei, in Bum 1932, p. 82. This is not primarily an anticlerical view, but is rather a metaphysical commitment involving God and his relation to the universe and to human beings. See esp. Garber 1993, p. 292; Part I of Voss 1993; and Rodis-Lewis 1993.

2.3. Update: Type/Law Convergent Realism

We can imagine many components of the ontological tyranny appearing in philosophical discussions of objective knowledge, truth, and reality, stripped of any overtly religious content. In fact, the assumption that *real* objectivity will result in a convergence on One True Description seems to be widespread.[26] In the words of the popular grisly philosophical metaphor, real knowledge "carves Nature at its joints." Obviously, this view of knowledge presupposes that Nature *has* joints, that is, "natural" objects and/or events, and kinds, and laws, which could serve (ideally) to guide inquirers, in a converging fashion, toward the discovery of these natural divisions. Under this view, which I call *type/law convergent realism*, the criterion of success for real knowledge is rather strict: it is not enough that Reality never has or never will *resist* our use of certain sets of categories of objects, events, kinds, or laws – if these sets don't conform to *nature's* categories, individuals, and laws, then we do not have *true* descriptions of reality. This is really just an updated version of the *ontological tyranny*. Let us now explore the views of two philosophers who might be thought, *prima facie*, to adhere to the above position.

2.3.1. BERNARD WILLIAMS'S ABSOLUTE CONCEPTION One current picture of "scientific objectivity" that is especially influential among moral theorists is presented by Bernard Williams. For him, what is distinctive about *scientific* inquiry is that scientific objectivity can lead – in principle – to a *convergence* on "the absolute conception" of Reality. The absolute conception is, he writes, "a conception of reality as it is independently of our thought, and to which all representations of reality can be related" (1978, p. 211). The independent existence of the relevant part of reality is central to Williams' claims about scientific objectivity and its relation to the absolute conception: "If knowledge is what it claims to be, then it

[26] E.g., Mackie 1976, p. 20; Nagel 1986, p. 83; Novick 1988. Cf. Putnam, 1981, pp. 49–55, 73. This requirement is rejected by some, on the grounds that we cannot make sense of such a notion of what we are supposed to be converging on. Putnam, for example, equates "the doctrine that all 'external questions' are without cognitive sense," with "the doctrine that no rational reconstruction is uniquely correct or corresponds to the way things 'really are'" (1981, p. 112). Here, Putnam conflates the issue of "making sense of statements about the world" with the issue of *convergence*. Although Putnam cites Carnap in his support, Carnap actually believed that there were any number of legitimate bases for preferring one reconstruction over another, we can sensibly approach questions about "the world" – what Carnap rejected was the ideal of *convergence* (1963, pp. 869–873; Richard C. Jeffrey, personal communication).

is knowledge of a reality which exists independently of that knowledge, and ... independently of any thought or experience" (1978, p. 64).[27] For Williams, the goal embodied in scientific inquiry is to reflect "on the world that there is *anyway*, independent of our own experience" (1985, p. 138).[28]

According to Williams, scientific inquiry is supposed to go something like this: ideally, scientific knowledge of reality, that is, "the absolute conception," is nonperspectival, and completely detached. As we increase "objectivity," which is understood here as a *method of detachment from any localized perspective combined with third-personal public access* to phenomena,[29] we are promised (in principle), to converge on greater and greater portions of the absolute conception. That is, we will have converging access to more and more of the Really Real.

In sum, the application of "objectivity" – as method, a set of standards of inquiry – will lead to "objective" knowledge, where "objective" knowledge is knowledge of the Really Real *"sub specie aeternitatis,"* that is, *part of* the ontology of the (objectively) Real (1985, pp. 111, 132, 136; cf. Smart 1963, p. 84).[30] Hence, under Williams's view, several types of objectivity are *necessary for and characteristic of* scientific inquiry. And Williams is not idiosyncratic.[31]

[27] Searle places similar emphasis on a type of "independence of reality" (1992, p. 192; 1993, p. 66).

[28] Hilary Putnam's recent claims, that there is no sense to be made of such a picture, seem to be based on a profound misreading of Williams's views (Putnam, 1981, pp. 49–55, 73–74, 1992, pp. 101–103, 108). For example, he takes Williams to be arguing that the absolute conception will contain only physics (Putnam 1992, pp. 83–85, 95, 99–100, 102, 107, 108). On the contrary, Williams says nothing about the ontological level or nature of the entities in the absolute conception that would commit him to a physical reductionist view. Putnam's remarks suggest that he has confused one of his own earlier incarnations with Williams (Putnam 1992, pp. 2–3). Also, much of Putnam's evident confusion is generated by his reading Williams's "absolute conception" discussion *as an argument*, rather than as the description it is *meant* to be (Putnam 1992, pp. 82–83, 92, 97, 100–102, 107).

[29] Note that this is the stronger version of "public accessibility" mentioned in Section 1. Specifically, "third-person accessible" does not include "common perceptual experience," that is, perceptual awareness that individuals have when exposed to the same phenomena (Williams 1978, p. 302).

[30] Even if we remain agnostic – as Williams does – about the possibility of anyone ever reaching the absolute conception, we are still promised *increased and converging access* to *independently existing reality*. Note that even if we acknowledge that no one can ever be *completely* detached, the degree of detachment is portrayed as positively correlating with increased and converging access to the Independently Existing part of the Really Real. See Williams 1985, pp. 111, 132, 136; cf. Smart 1963, p. 84.

[31] Antony 1993; Boyd 1980; Dennett 1978a, 1978b, p. 256, 1991, esp. p. 71; Hesse 1974, 1988, pp. 234–236; Mackie 1976, pp. 18–19; McMullin 1988a, pp. 35, 39–44, 1988b, pp. 233–224;

In spite of Putnam, Rorty, and others casting Williams's position as an updated version of the ontological tyranny, this is not correct.[32] Williams is talking only about *scientific knowledge*; he's simply arguing that only third-person accessible and ontologically independent aspects of reality are ultimately going to be successful objects of *scientific* inquiry, and only *they* can be expected to appear in the absolute conception. In fact, he explicitly excludes our knowledge of "common perceptual experience" and "social science" from the absolute conception. The claim is that the absolute conception can "make sense of how natural science can be absolute knowledge of reality," even though it is not complete – it does not include *everything* that is real (1978, p. 302; 1985, pp. 111, 135–140, 148).[33]

By excluding everything peculiar to us as human beings, or as individuals, or as social beings, Williams very deliberately opens the door to different sorts of investigations which do take these aspects of reality into account; in fact, he advocates more local and contextual analyses and evaluations of the sorts of things – which presuppose interests and practices of a "social world" – that do not figure directly in the absolute conception (1985). Hence, Williams's views do not have the capacity to eliminate certain contextual factors – such as sex and gender – as relevant, without argument.

Perhaps Williams would say that the important thing is that sex and gender are irrelevant to the absolute conception, although they're not irrelevant categories for social sciences, anthropology, history, and psychology. Still, Williams admits that the areas in which *he* works involve social, anthropological, and psychological factors and descriptions, yet he has ignored a central one without a peep.

2.3.2. CHARLES PEIRCE'S REALISM Charles S. Peirce is often characterized as an archetypal convergent realist; he is cast as presenting a modernized,

Nagel 1980, pp. 78–79, 1986, pp. 4–5, 16–17, 77, 83, 1993, pp. 37–38; Novick 1988; Quine 1981, pp. 92, 98; Rescher 1992, pp. 12, 14; Scheffler 1967, pp. 9–10, 13–14; Shapere 1981; Smart 1963, p. 84; and Wiggins 1976, p. 360, make similar claims, as do many others.

32 Putnam claims: "Williams believes that any conceivable species of intelligent beings (if they frame hypotheses properly, perform the appropriate experiments) can 'converge' toward agreement on the laws of the ideal physics, in the fashion first envisaged by C. S. Peirce" (1992, p. 84; Putnam cites Peirce's 1877 in support of this claim). See also Rorty 1982, p. 190.

33 Cf. Williams 1978, pp. 245, 301. Putnam seems to have missed this vital facet of Williams's view, when he writes about "the absolute conception (which is equated with the complete description of the world in terms of Primary Qualities) . . . " (Putnam 1992, p. 98, also pp. 84–85, 92–94, 102). Williams put it in black and white: "It is centrally important that these ideas relate to science, not to all kinds of knowledge" of reality (1985, p. 139).

secular version of the ontological tyranny. Wiggins, for example, describes a "Peircean view of Science . . . as discovering that which, the world being what it is, is destined to be ultimately agreed to by all who investigate" (1976, p. 361). In support, Wiggins quotes from Peirce's 1878 paper, "How to Make Our Ideas Clear": "The opinion which is fated to be ultimately agreed to by all who investigate, is what we mean by truth, and the object represented in this opinion is the real" (1878a, p. 139).[34] Wiggins's misunderstanding of Peirce's view is revealed in his interpretation of the "all" in the above sentence; Wiggins writes, "Let 'all' mean 'all actual or possible intelligent beings competent, whatever their conceptual scheme, to look for the fundamentally explanatory principles of the world" (Wiggins 1976, p. 361). As it stands, Wiggins's claim is ambiguous about whether the ultimate conceptual scheme is *necessary*, but he subsequently attributes to Peirce the view that "there is a reality which dictates the way a scientific theory has to be in order that what happens in the world be explained by the theory" (1976, p. 362; my emphasis).

Wiggins claims further that Peirce's "fundamentally explanatory principles" will include: "the real properties of the world, the properties which inhere in the world *however it is viewed*, [which] are the primary qualities" (1976, p. 362). Ultimately, Wiggins characterizes Peirce's view as "the 'external' perspective," to which it is absurd to try to add on some component of "human commitment." Wiggins concludes: "This Peircean conceptual scheme *articulates* nothing which it is humanly possible to care about" (1976, p. 363).

As we shall see later, Peirce's scheme actually includes nothing that human beings *do not* care about. I would now like to emphasize several aspects of Peirce's thought that clearly disqualify him as a *type/law convergent realist*, despite the repeated attribution to him of this view.

INDEPENDENCE AND REALITY The quotes selected by Wiggins and others, such as Shimony, Rorty, Murphy, and Nagel, with their appeal to "fated," ends, may appear to commit Peirce to a "Truth-will-out" view of knowledge and reality, in which Real things eventually force inquirers into True understandings of them.[35] But this is the exact opposite of

[34] Reprinted in Peirce 1992. References to Peirce's papers will appear with their original dates of publication, while page numbers will refer to the 1992 edition, unless otherwise noted.

[35] For instance, Shimony attributes to Peirce "an infallible asymptotic approach," and Rorty objects to Peirce's "view 'that truth is fated to win'" (Shimony 1970, p. 127; Rorty 1982,

Peirce's view. Immediately following the widely-quoted passage regarding the "opinion . . . fated to be ultimately agreed to . . . ," Peirce insists that his view "makes the characters of the real" depend on "what is ultimately thought about them" (1878a, p. 139). Peirce argues that his view of Reality is therefore incompatible with what he calls an a priori or "abstract" definition of reality (which is fundamentally equivalent to the ontological tyranny), according to which "we . . . define the real as that whose characters are independent of what anybody may think them to be" (1878a, p. 137). Even more damaging, he concludes that Reality – what anything really is – "depends on the ultimate decision of the community" (1868, p. 54).

This dependence of the Real on our thoughts about it is clarified by Peirce in the next section; Reality is not to be understood as independent of thought in general, but as independent "only of what you or I or any finite number of men think about it" (1878a, p. 139). And the "opinion which would finally result from investigation does not depend on how anybody may actually think. . . . But the reality of that which is real *does depend* on the real fact that investigation is destined to lead, at last, if continued long enough, to a belief in it" (1878a, p. 139; my emphasis; cf. Putnam 1975, p. 272). Hence, in spite of this genuine dependence on community-mediated cognitions, the outcome of such investigation "is the real, as it really is" (Peirce 1868, p. 52; *see* Hacking 1983, p. 58). Given Peirce's views regarding the essentially social nature of inquiry, his commitment to the dependence of the Real on inquirers packs a real punch; it is extremely difficult to reconcile with any interpretation of *type/law convergent realism.*

Perhaps Peirce's decisive break with *type/law convergent realism* is made most clearly in his views about Types and Laws of Nature themselves. Briefly, he thinks that Laws of Nature are those among an infinity of regularities in the universe, distinguished only by the fact that *we* are interested in them; there is nothing "inevitable" about them, or about *their appearance* in the final results of inquiry. Here is Peirce describing his own views on regularities of nature from his 1878 paper "The Order of Nature": "I remarked there that the degree to which nature seems to present a general regularity depends upon the fact that the regularities in it are of interest and importance to us, while the irregularities are without practical use or significance" (1900–1901 (1958, p. 114)). He also rejects

pp. xiv, 173; also Murphy 1990, p. 30; Nagel 1986, p. 83; Rescher 1978, pp. 1–10, 16, 95; van Fraassen 1989, pp. 21–23).

the assumption or belief "that every single fact in the universe is precisely determined by law" (1892, p. 298).

Although Peirce calls himself a "realist" about Types, his is not the sort of realism in which Types correspond to some sort of "Natural Kinds" or natural divisions in Nature – and this lack of correspondence results not simply because of the fallibility inherent in any stage of inquiry; it also holds for the Types on which the ideal community of inquirers would ultimately agree.

On the status of Laws and Types in Nature, Peirce starts from the logical point that "any plurality or lot of objects whatever have some character in common (no matter how insignificant) which is peculiar to them and not shared by anything else" (1878b, p. 174). This means that there are infinitely many ways of dividing the universe up into Types, or of grouping things together by *similarity*. He further argued that we ought to consider the characters of things "as relative to the perceptions and active powers of living beings" (1878b, p. 175). He concludes from this that if we do not rank characters by relative importance, there would be no greater or lesser degree of uniformity in the world (1878b, p. 175); hence, we have to decide which characters to focus on, in order to make any inductive generalizations or Laws of Nature at all (1878b, p. 176).

Peirce does claim that "the real is that which is such as it is regardless of how it is, at any time, thought to be" (1905b [1955, p. 301]). This could be interpreted as a commitment to the sort of a priori realism that Peirce claimed to oppose; but Peirce insists on a very specific interpretation of what 'real objects and their characters' means, here: "And in what does [the behaviour of different possible kinds of material substance] consist except that if a substance of a certain kind should be exposed to an agency of a certain kind, a certain kind of sensible result *would* ensue, according to our experiences hitherto. As for the pragmaticist, it is precisely his position that nothing else than this can be so much as *meant* by saying that an object possesses a character" (1905b [1955, p. 301]; his emphasis).

Hence, we must interpret the following widely-cited claim of Peirce's very carefully: "That whose characters are independent of how you or I think is an external reality" (1878a, p. 136). He clarifies the fact that he does not identify reality with independence from us, in his discussion of phenomena "within our own minds, dependent upon our thought, which are at the same time *real* in the sense that we really think them" (1878a, p. 136; my emphasis). Even though the characters of these phenomena in the mind depend on how we think, "they do not depend on what we think those characters to be"; hence, Peirce concludes that "a dream has a real existence as a mental phenomenon," but the contents of the dream don't depend on what anybody thinks was dreamt – the contents are "independent of all opinion on the subject" (1878a, p. 137). Peirce then defines as *the real*: "that whose characters are independent of what anybody may think them to be" (1878a, p. 137). The crux of Peirce's view is that he separates the dependence of the *existence and properties* of certain phenomena from the issue of *what we believe* about those properties and phenomena.

Peirce's view is further clarified by his attack on the "assumption" that "what is relative to thought cannot be real." Peirce asks: "But why not, exactly? *Red* is relative to sight, but the fact that this or that is in that relation to vision that we call being red is not itself relative to sight; it is a real fact" (1905a [1955, p. 264]). Note that Peirce's position on "subjective" things – including, in this case, secondary qualities – is precisely the opposite of Wiggins's ascription, quoted earlier.

Peirce's views about individuation and categorization of Types are also essential to his views about our inductive grasp of Laws of Nature. Peirce insists, versus John Stuart Mill, on the dependence of Laws of Nature on our picking out of characters that are important to us (1878b, p. 179).[36]

Furthermore, Peirce thinks that certain conditions are "essential to the validity of inductions" – which, after all, relies on determination of similarities in conditions, causes, and effects (1878b, p. 179):

> When we take all the characters into account, any pair of objects resemble one another in just as many particulars as any other pair. If we limit ourselves to such characters as have for us any importance, interest, or obviousness, then a synthetic conclusion may be drawn. (1878b, p. 179)

That is, the similarities which are essential to any sort of induction depend on our interests. As Peirce himself said, reflecting late in his career: "the most striking feature of [pragmatism] was its recognition of an inseparable connection between *rational cognition and rational purpose*" (1905a [1955, p. 253]; my emphasis).

In sum, the popular representations of Peirce's views on the inevitability of convergence on one Truth are essentially incomplete, because they fail to address the deep contingency and interest and value-dependence involved in his understanding of that Truth; in other words, Peirce's views

[36] Part of this argument is an attack on Mill's statistical methods; Peirce's method demands that we designate the character(s) under investigation *before* any statistical samples are taken, in order for the resulting data to count as a test of a Law or generalization.

stand in direct opposition to the *ontological tyranny*. Peirce insisted on the irreducible and necessary dependence of Types and Laws of Nature on the interests of scientific communities; I conclude that Peirce's views cannot be used to rule out, a priori, any examination or critique of those interests on the basis of sex and gender.

2.4. Conclusions

In summary so far, I have emphasized, in discussing the *ontological tyranny*, the historical context and motivations of the original distinction between Primary and Secondary Qualities, because it is extremely important that this configuration of ontological and methodological views constituted a *particular formulation of standards of "objectivity" for the physical sciences* that was, in turn, dependent on religious beliefs. Next, I examined the views of two philosophers prominently cited as holding nonreligious and very strict standards of objectivity and knowledge. I argued that neither Williams nor Peirce can, within their own systems, rule out the potential relevance of a variety of social considerations – including those of sex and gender – even in our most strict, scientific, and respected inquiry. Indeed, Williams's limiting of potential objects of objective convergent knowledge, and Peirce's explicit and radical insistence on the roles of human interests pervading all aspects of inquiry, seem to invite the inclusion of biological, psychological, and social factors into our philosophical analyses of knowledge, truth, objectivity, and reality. This may be somewhat surprising, because Williams and Peirce have so often been set up as opponents by those currently advocating the indispensability of the social, lived human context in our philosophical understanding of language, meaning, truth, and objectivity. We turn now to a sample of these thinkers.

3. VARIANTS OF OBJECTIVITY

There are many ways of redefining and connecting meanings of *objectivity* that reject the *ontological tyranny*.[37] Such revisions usually involve

[37] Vocal opponents of the ontological tyranny have come from all quarters: e.g., Cartwright 1983, 1990; Dupre 1983, 1993; Fine 1984, 1986; Hacking 1983, 1988; Hull 1988; Johnson 1987; Laudan 1981; McDowell 1979, pp. 347, 346, 341, 1988; Polanyi 1969a; Suppes 1978; Wiggins 1976, pp. 338, 340, 363; and much of the literature of feminist epistemologists, esp.: Bordo 1987, 1989; Code 1991, 1993; Haraway 1989, 1991; Haslanger 1993; Jaggar and Bordo 1989; Rose 1983; and Whitbeck 1984.

refinements of "objective methods" or limitations on the applicability of objective methodology for knowledge of all aspects of Reality. I would like to summarize, briefly, which parts of the *ontological tyranny* are dropped or transformed in the refinements of the notion of *objectivity* offered by four diverse authors: Rudolf Carnap, John McDowell, Thomas Nagel, and John Searle. I will argue that none of these authors, or those with similar positions, are in a position to exclude the potential relevance of analyses of sex and gender from their understandings of knowledge, meaning, and truth, without argument. In fact, according to their own arguments and standards, such analyses ought to be taken as central.

3.1. Rudolf Carnap

Carnap's many discussions concerning how to make choices among different languages are sometimes spurned on account of their dismissal of "real" philosophical problems as "pseudo-problems." This quibble aside, I would like to illustrate briefly how Carnap's views about the essential sociality of science and its relations to our community's purposes and goals, invite the examination and discussion of sex and gender, among many other facets of social life.

As Carnap describes himself in 1963, "I put now more emphasis than previously upon the social factor in both the acquisition and application of knowledge, be it common sense knowledge or science; furthermore, upon points where the development of a conceptual system or of a theory involves practical decisions; and upon the fact that all knowledge begins with and serves the relations between a living organism and its environment. It is certainly important to keep these aspects in view in order to fully understand such social phenomena as language and science" (1963, p. 861).[38] And Carnap, in a passage remarkably prescient of Longino's argument nearly thirty years later, emphasizes the philosophical centrality of problems concerning choices "among conceptual frameworks... both of theoretical investigations and of practical deliberations and decisions with respect to an acceptance or a change of frameworks, especially of the most general frameworks containing categorial concepts which are fundamental for the representation of all knowledge" (1963, p. 862).

Fundamentally, Carnap thought that reasoned discussion, in terms of the purposes and goals set by the parties in question, was possible and

[38] Note that these concerns were not new for Carnap; see his 1928, 1950, 1956, 1967.

desirable, when considering different or untranslatable languages concerning a particular topic. Carnap gives an example of how such a negotiation about languages could work: "They proceed to communicate to each other their preferences and practical decisions concerning the acceptance of the two languages and their reasons for the decisions" (1963, p. 868, cf. pp. 862–869).

In sum, the irreducible presence of preferences and practical decisions, made within communities of inquirers in a reasoned fashion, clearly allows for concerns about all sorts of issues, including the sex and/or gender structures or coding of languages or conceptual schemes, to count as legitimate topics. This position has deep similarities to Peirce's, which Carnap himself acknowledged; the basic move was to transfer discussion from a priori intuition-bashing to interactive and concrete discussions of the consequences, goals, and reasons involved in seeing things one way rather than another.

3.2. John McDowell

John McDowell also rejects the ontological tyranny, and offers a reworked definition of *objectivity*. He rejects the requirement of *detachment* for objectivity, and focuses on discussion of the *publicity* – not of the phenomena in question – but of the standards of judgment accepted and embodied in community practices (1979, pp. 339–341).

McDowell's abandonment of the requirement of detachment is founded on his rejection of the specific connection, assumed in the *ontological tyranny*, between the independent existence of the phenomena being investigated – most especially, on their independence *from us* – and the methodological value of *detachment*. McDowell explicitly bases his rejection of the utility and desirability of detachment on Wittgenstein's conclusions that even mathematical truths are not independent from us and our practices (McDowell 1979, p. 341; 1988, p. 170).

The alternative McDowell offers for detachment is best illuminated in his discussion of our "immersion" in forms of life, and the role of human communities and standards of judgment in knowledge, truth, and meaning. Ultimately, McDowell defends what he identifies as a Wittgensteinian view, that we do not have to postulate a psychological mechanism (or rule) underlying behavior in order to *understand* someone doing something correctly (1979, p. 337). The question immediately arises as to the ground and nature of our "confident expectation" that someone will perform appropriately. Here, McDowell turns to Stanley Cavell's discussion about "the competent use of words" as a model:

> ... we learn and teach words in certain contexts, and then we are expected, and expect others, to be able to project them into further contexts. ... That on the whole we do [this] is a matter of our sharing routes of interest and feeling, modes of response, senses of humour and of significance and of fulfillment, of what is outrageous, of what is similar to what else, what a rebuke, what forgiveness, of when an utterance is an assertion, when an appeal, when an explanation – all the whirl of organism Wittgenstein calls 'forms of life'. Human speech and activity, sanity and community, rest upon nothing more, but nothing less, than this. (Cavell 1969, p. 52)

McDowell takes from Cavell the lesson that "it is only because of our own involvement in our "whirl of organism" that we can understand the words we produce as conferring "that special compellingness on the judgment explained" (McDowell 1979, p. 339). In fact, McDowell claims, even the paradigm cases of rationality all have "dependence on our partially shared 'whirl of organism'" (1979, p. 341). Hence, we cannot recognize reason itself from outside the practices of a given community; and this conclusion applies to all cases of reasoning, including deductive argument, and not merely to reasoning about morality or virtue (1979, pp. 345–346).

In the end, McDowell rejects the idea that scientific method gives us a more external or more objective viewpoint; he locates *objectivity* solely within a given conceptual framework, and concludes that we should "give up the idea that philosophical thought, about the sorts of practice in question, should be undertaken at some external standpoint, *outside our immersion in our familiar forms of life*" (1979, p. 341; my emphasis). McDowell favors a very lean version of "objectivity" that amounts, essentially, to a willingness to submit one's practices to public, accepted community standards of concepts and frameworks. In rejecting the legitimacy of a "neutral external standpoint," McDowell maintains that what is publicly shared is the "conceptual equipment which forms the framework" within which we conceive the world, and that 'objectivity' is properly conceived only *within* this conceptual framework (1979, pp. 345, 347).

I would like to offer two observations at this point. First, note the strong resemblance between McDowell's views and Carnap's – they both see certain fundamental evaluations – of truth, or of objectivity – as legitimate only *within* certain languages, conceptual schemes, or social contexts. Carnap, though, explicitly addresses the issue of how to choose among various possible languages or frameworks, and he believes that these choices are properly made within community discussions of goals, purposes, and reasons. Carnap's vision of cooperative and rational engagement within a community regarding the *ends*, as well as the reasonableness of the means, of conceptual schemas or languages brings the picture of objectivity and rationality up to a meta-level that is comparatively neglected by other authors. One important recent exception is Helen Longino's detailed examination of *how and why* critical discussions of the goals, frameworks, and standards of scientific inquiry are possible and desirable. Longino's analysis focuses on explicitly *feminist* critiques of scientific inquiry, practices, and standards – but her analysis bears strong similarities to the views of Peirce and of Dewey, and it extends effectively to any set of community goals and standards.

The second point is that this continuity between pragmatic and explicitly "feminist" approaches to scientific knowledge is extremely revealing; it implies that those authors who commit themselves to taking the interests, values, and goals of the relevant communities into account are obliged, on the face of it, to attend to the influences within those communities of the categories of sex and gender. One might respond: listen, we can't take everything about the community's values and standards into account – and this is, of course, exactly right. But given that there is virtual unanimity among anthropologists that sex and gender roles lay the foundations of every human society's other social practices – including communication, lines of authority, distribution of physical, emotional, and intellectual goods, and the very general social structures of who decides what – the burden of proof seems to rest squarely on those who want to *include* these more superficial social practices as vital to philosophical understandings of meaning, truth, language, and knowledge-acquisition (e.g., McDowell and Wiggins[39]), while they *exclude* the bedrock social roles of male and female, masculine and feminine – on which these other social practices are overlaid and constructed.

In the particular case we've just examined, I find it highly peculiar that we have McDowell and Cavell including everything but the kitchen sink as relevant to this "whirl of organism" and our conceptions of objectivity, truth, knowledge, and meaning – but not sex and gender.

[39] Wiggins's views are, for the most part, similar in content and motivation to McDowell's. See Wiggins 1976, pp. 341, 370–371, 1987, esp. pp. 174–175, 1991.

3.3. *Thomas Nagel*

Thomas Nagel departs from the *ontological tyranny* in two crucial ways: (1) he abandons the demand for the third-person accessibility of the phenomena; and (2) the requirement that the phenomena be independently existing from us as knowers – from which we are "detached" – is remodeled to allow us *as valuers* to be a distinct subset of the subjective part of reality.

The situation is this: our experiences, in their irreducible "what it's like"–ness, do not fulfill the "objective" methodological requirements of the *ontological tyranny*. And the reason that this aspect of our experiences doesn't fulfill the methodological requirements is *ontological* – it has to do with the very nature of those experiences. The crucial trait here is that they are attached, from a particular point of view. The fact is that we don't seem to be able to *separate out* this trait of attachment from our experience, without destroying it in some way.

My reading of Nagel (and also of John Searle) is that they are in the business of arguing that things that don't exist independently from us are not any less real in virtue of that dependence; I think that these are *clearly intended* as ontological claims. Searle, whose views I discuss in the next subsection, consistently keeps his focus on this defect of the ontological tyranny.

Nagel, in contrast, goes on about "the view from nowhere," which embodies a specific methodological program. In his book, *The View From Nowhere*, Nagel argues that we can separate out our attachment from our experience. Nagel calls the process of progressive detachment – of surgically removing parts of our experience that are peculiar to our own point of view – the "ascent to an objective view."[40] He offers a modified form of "objectivity," which expands a method of detachment into the subjective realm of reality, where it enables us to construct an "objective self."[41] Although he acknowledges that increased objectivity – in the sense of detachment – cannot provide a complete picture of the world, he also states that objectivity (as method) is *the only* path to knowledge of reality.[42]

Hence, although Nagel emphasizes the methodological value of detachment, he also insists that there are still parts of reality unreachable

[40] 1986, pp. 140–141; *see* 1979, p. 206, 1980, pp. 79–81, 91.
[41] 1986, pp. 62–66; cf. 1980, pp. 81–83, 118.
[42] 1979, p.212, 1980, p. 91, 1986, p. 26.

even by his revised "objective" method; he urges that, pragmatically, we must recognize and deal with its limitations (1986, p. 7; 1979, p. 213). It is ironic that Nagel has argued forcefully for the reality of the subjective, yet he insists here that we can have little knowledge of it. Among the things that will be left out by application of his methodological objectivity to the conception of mind will be: "the exact character of each of the experiential and intentional perspectives with which it deals," which "can be understood only from within or by subjective imagination."[43]

The fact that Nagel ultimately requires some form of detachment for "objectivity," while he does *not* require third-personal publicity of the phenomena in question, is problematic. One question arises immediately: without publicly accessible phenomena – either in the weak sense of humanly obtainable experience, or in the strong sense of third-person cross-checking – how does his "reflective process" of the "ascent to objectivity" amount to more than "mere intersubjectivity"?

Nagel responds by insisting that "even subjective concepts" are characterized by a certain, limited *objectivity*: "Just as adrenalin would exist even if no one had ever thought about it, so conscious mental states and persisting selves could exist even if the concepts didn't . . . [but] what *are* these things, apart from the concepts which enable us to refer to them?" (1986, p. 35). Nagel concludes that "there must be a notion of objectivity which applies to the self, to phenomenological qualities, and to other mental categories," because the idea of making a *mistake* about these things *makes sense*; that is, "there is a distinction between appearance and reality in this domain" (1986, p. 36).

The way that we "make sense" of the appearance/reality distinction, even with the most subjective or "private" phenomena, is through community-wide practices or conventions that are acquired and committed to. Nagel bases this conclusion on his interpretation of Wittgenstein. According to Nagel, "even our most subjective phenomenological concepts are public in a sense . . . since a kind of intersubjective agreement characterizes even what is most subjective [as Wittgenstein showed], the transition to a more objective viewpoint is not accomplished merely through intersubjective agreement."[44] That is, the intersubjective adoption of community standards doesn't necessarily get us "outside" of the world of subjective experience. What, then, is the argument for the claim that the "idea of objectivity always points *beyond* mere intersubjective

[43] 1980, p. 90; cf. p. 118; also 1979, p. 170.

[44] Nagel 1979, pp. 207–208; cf. Wittgenstein 1953, 1965; McDowell 1979.

agreement even though such agreement, criticism, and justification are essential methods of reaching an objective view" (Nagel 1986, p. 108; my emphasis)? Nagel's answer is that the whole point of objectivity is "to talk about the world itself" (1986, p. 108), and he rejects that there are grounds for drawing Wittgenstein's distinction between legitimate and illegitimate extensions, beyond the range of actual agreement in judgments; the external world "is not dependent on our view of it, or any other view: the direction of dependence is the reverse" (1986, pp. 108–109). I take it that Nagel is emphasizing that, while social and contextual interrelations are necessarily involved in meaning, truth, and knowledge, they are not determinative.[45] Still, Nagel's acknowledgment that these factors play the central roles they do obliges him to explain why sex and gender categories are not among the social and contextual factors that count.

3.4. John Searle

Nagel's position looks weak and confused in comparison to John Searle's rejection of anachronistic standards for science and knowledge. Searle makes the genuinely interesting move: he rejects the ontological tyranny as appropriate to scientific knowledge – especially of the mind/brain – and thereby undermines the legitimacy of excluding subjective and attached phenomena from our ontology at the start (1984, 1992).[46]

More specifically, by rejecting the appropriateness of an ontologically loaded requirement of *detachment* – with its assumptions of the complete independence of the Really Real – Searle undermines the traditional *ontological tyranny* in a fundamental way. Because we *do* have every reason to ascribe full-fledged Reality to our subjective or attached aspects of experience, there must be something wrong with the *specific standard of science itself* which requires detachment. "If," Searle writes, "the fact of subjectivity runs counter to a certain definition of 'science,' then it is the definition and not the fact which we will have to abandon" (1984, p. 25). Not surprisingly, his vision of a science of the mind/brain is in danger of being branded as "unscientific" by many philosophers and scientists who are already committed to the linkages among the ontological tyranny, "objectivity," and science itself. Even Nagel, who holds

[45] Note that Nagel's view of the dependence of reality on our descriptions of it is the reverse of Peirce's actual view.

[46] As Searle puts the point in *Mind, Brains and Science*, "there *really are* mental states; some of them are conscious; many have intentionality; they all have subjectivity . . . " (1984, p. 27; my emphasis).

similar interests in subjective states, allows his unexamined assumption of the legitimacy of the ontological tyranny to prompt doubts concerning Searle's entire project; Nagel steadfastly denies that any change of the character Searle requires could take place *within science*.[47] Nagel's pessimism is very revealing: it shows his unwillingness to challenge the central ties binding together the ontological tyranny, and it also demonstrates that he has a very specific and anachronistic picture of science itself; at the same time, it highlights the radical nature of Searle's ontological views.

Having dispensed with the *ontological tyranny*, Searle's ontology includes both subjective and objective realities (1992, p. 99). But he still needs to address how these realities relate, and Searle does so through his descriptions of intentional content, or how we get the meanings we do. His basic claim is that our meanings rely on what he calls *the Background*, which includes a broad range of human practices: "...each sentence is interpreted against a Background of human capacities (abilities to engage in certain practices, know-how, ways of doing things, etc.), and those capacities will fix different interpretations, even [when] the literal meaning of the expression remains constant" (1992, p. 179). In other words, "Sentence meaning *radically* underdetermines the content of what is said" (1992, p. 181).

The dependence of our meanings and directed states of awareness on this essential Background runs very deep, according to Searle:

> ... the Background concerns not merely such relatively sophisticated problems as the interpretation of sentences but such fundamental features as those that constitute the formal basis of all language.... The syntactical unit [in some cases]... is not a word in the traditional sense, but a sequence of token inscriptions. Similarly with the systems of opposition that the structuralists were so fond of: The apparatus of hot as opposed to cold, North to South, *male to female*, life to death, East to West, up to down, etc., are all Background based. *There is nothing inevitable about accepting these oppositions*. (1992, pp. 185–186; my emphasis)

The fact that Searle explicitly acknowledges the contingency of certain features of the Background here, including sex differences, leads to a serious problem with his dismissal of the intellectual relevance of feminist thought. The problem is related to a deep problem with Searle's notion of Background itself. Sometimes, it seems that he thinks of the Background

[47] Nagel 1993, pp. 39–41; see P. S. Churchland 1993, pp. 30–31, 48.

as a lump, that is, as undifferentiated and diffuse; other times, he individuates parts of the Background, in order to discuss their relevance to specific cases. This ambiguity in the concept of Background plays an important role in Searle's attitudes about the potential relevance of Background assumptions of intentional states involved in scientific knowledge, beliefs, and judgments. In particular, Searle's argumentative strategies and standards for discussing the status of undifferentiated versus individuated Background are not applied consistently.

Searle begins with the claim (1) that the Background underlies all intentional states. In discussing how the Background relates to specific intentional states, Searle also claims (2) that some parts of the Background are irrelevant to some specific intentional states. In order to demonstrate claim (2), Searle offers the example of the role of Background skills involved in being able to order a hamburger, and the intentional states involved in solving a math problem; his claim is that the hamburger-ordering part of the Background is irrelevant to the specific intentional states of doing the math. This is very plausible; but the crucial point is that Searle, in defending his general claim (2), of the irrelevance of some parts of the Background to some intentional states, offers case-by-case arguments involving particular examples.

Now, consider Searle's claim (3) (quoted above) that sex differences – or the polarity between male and female[48] – are a ubiquitous part of the Background. Nevertheless, Searle also claims (4) that they are irrelevant to most intentional states involved in real science. Searle claims: "the objective truth or falsity of [knowledge] claims made is totally independent of the motives, the morality, or even the gender, the race, or the ethnicity of the maker" (1993, p. 66). This is in spite of his claim, two pages later, that "even objectivity only functions relative to a shared 'background' of cognitive capacities and hence is, in a sense, a form of intersubjectivity" (1993, p. 68).[49]

Here is the problem: Searle has set a standard, in his own exposition, of providing cases and examples in order to demonstrate the irrelevance of certain parts of the Background to particular intentional states. Yet he offers none to support his conclusions about the irrelevance of Background assumptions of sex or gender differences to intentional states

[48] It does seem that Searle also intends gender, that is, the polarity between masculine and feminine, to be included here, but this is not necessary to my argument.

[49] Oddly, Searle later attacks those who claim that objectivity and universality "tend to reflect local historical conditions" (1993, p. 69). It is difficult to reconcile his rejection of this claim with his own descriptions of the Background.

involving scientific reasoning; in fact, he *dismisses* the potential relevance of these parts of the Background out of hand, without argument. Worse yet, he cannot claim that it is illegitimate to question, challenge, or examine the presence of the male/female dichotomy in the Background, because he explicitly claims that such a distinction is eliminable.

Hence, Searle has not only failed to eliminate the potential relevance of sex differences for understanding our meanings, scientific knowledge, and intentional states, he has actually given an argument supporting the *prima facie* centrality of questions about sex and gender for these understandings.

3.5. Conclusions

My conclusions for this section are as follows. Notice, first, that a *consensus* about the appropriate way to redefine "objectivity" – once the ontological tyranny is rejected – is glaring in its absence.

Furthermore, the take-home message of all of the above variants of "objectivity" (except Searle's) is that the concept is community-based or socially grounded in its significance to knowledge and truth. And Searle explicitly claims that the social conventions and practices, including those involving sex and gender, criss-cross the entire background – the keystone of his theory of intentionality. Hence, according to their own standards, all of these authors ought to allow sex and gender as potentially relevant dimensions of the complex contexts that *they* see as necessary to understanding *objectivity* and *objective knowledge*.[50]

4. TRUTH OR CONSEQUENCES

In this section, I argue that once *type/law convergent realism* is abandoned, sociocultural issues arise immediately; in addition to resistances by reality, values and interests, broadly speaking, are necessarily involved in the

[50] Other, more direct investigations of the *social* dimension of judgments, inquiry, meaning, and truth have been most thoroughly pursued in feminist philosophy of science and epistemology. For example, see Alcoff and Potter 1993; Antony and Witt 1993; Babbitt 1993; Dupre 1993; Harding 1991, 1992, 1993; Haslanger 1993; Longino 1990, 1993a, 1993b; Nelson 1990, 1993; Rouse 1987; Tuana 1989; Wylie 1988, 1989. See also Bernstein 1983, 1988; Dewey 1925, 1929, 1938; Foucault 1972, 1987, pp. 96–98, 103, 107–112; Hacking 1988, p. 149; Haskell 1990, pp. 131–139; Hesse, 1974, 1988; Hull 1988; Johnson 1987, pp. 190–195, 200, 212; Kuhn, 1970, 1977, 1992, pp. 11, 13–15; Latour 1993; Nietzsche 1967; Peirce 1871, 1878a; Polanyi 1969a, b; Proctor 1991; Rorty 1988, p. 71; Scheffier 1967, pp. 1–10; Sellars 1968, esp. pp. 220–226; Wilson 1990; Ziman, 1968.

development of knowledge and concept-formation. One relevant set of interests is freely acknowledged: part of why we are interested in the phenomena we are, is because we are a specific kind of animal – this big, with these senses, needing these things, with these brains, living in these communities, with these aims, with this "whirl of organism," and so on, as reviewed in Section 3.

Take the belief that there really is some universe, some reality, independent of each and all human existences. Add the belief that we, as groups of human beings, families, nations, or specific kinds of animals, have ways of understanding this universe, this reality, ways that can vary across individuals, groups, time, or distance. Specifically, we may divide up the world into different types of things – different categories; and we may attend to different regularities of the universe, depending on our needs, interests, or values – that is, we may arrive at various laws of nature (as Peirce argued in his 1878b [1992, p. 15]).

At this point, most philosophers (and some scientists) will ask: Can we compare the relative merits of these different ways of dividing Reality up? Aren't some more "true to Reality" than others? Suppose that two groups have the customs of using very different sets of categories, and that the two groups seem to believe opposite truths about the same thing?

While I acknowledge that these are pressing questions, I wish to set them aside in this context. The important claim, for our purposes, is that – *absent* a commitment to type/law convergent realism – *it is possible* that there are different categories and laws that are good and true descriptions of the universe; some will want to limit their variation in such a fashion that they are all compatible, or intertranslatable, or explainable by some third point of view, or logically consistent. Nevertheless, the pivotal point here is the acknowledgment that Reality doesn't force ideal knowledge into a single, unique form, without some prior commitment to *type/law convergent realism*. The reason this acknowledgment is crucial is that any view of knowledge and reality that does *not* envision a unique and determinate knowledge of reality as its end, is subject – *legitimately and according to its own terms* – to feminist and other challenges that focus on the irreducible importance of social life, practices, and standards of judgment.

5. THE DOUBLE STANDARD

The belief that "feminist epistemology" is an oxymoron – because real knowledge of reality involves "objectivity," and "objectivity" *just means* "neutral," "nonideological," and "distanced from any personal interests or idiosyncrasies" – collapses under scrutiny.

It is "pure epistemology," "value-free inquiry," and "disinterested knowledge" that are the oxymorons. Mainstream contemporary epistemologists and metaphysicians acknowledge this in their own work, so why do they implicitly hold feminist philosophers to a different standard?[51] The emphasis on the limitations of objectivity, in specific guises and networks, has been a continuing theme of contemporary analytic philosophy for the past few decades. The popular sport of baiting feminist philosophers – into pointing to what's left out of objective knowledge, or into describing what methods, exactly, they would offer to replace the powerful "objective" methods grounding scientific knowledge – embodies a blatant double standard that has the effect of constantly putting feminist epistemologists on the defensive, on the fringes, and on the run.

This strategy can only work if "objectivity" is transparent, simple, stable, and clear in its meaning. It most certainly is not. In fact, taking "objectivity" as a sort of beautiful primitive, self-evident in its value, and all-powerful in its revelatory power, requires careless philosophy, and the best workers in metaphysics, epistemology, and philosophy of science have made reworked definitions of "objectivity" absolutely central to their own projects. Nevertheless, these redefinitions of objectivity and their relations to various sorts of knowledge often remain in the background, whereas the very visible set of views embodied in the ontological tyranny often serves as the default definition of "objectivity" that is rolled out when something authoritative and solid is sought – for example, when feminist critiques arise. Any appeal to a rejected but still default position on "objectivity" amounts to the strategy of "bait and switch," which is considered dishonest even in business, much less in philosophy.

In conclusion, there ought to be no such "ghetto" as "feminist epistemology."[52] Classic feminist concerns with exploring the impact of sex and gender on knowledge, understanding, and other relations between human beings and the rest of the world fall squarely within the sort of human and

[51] See n. 3. As an instance of the quantifiable social facts I refer to, I offer the following information about Sections 3.2–3.4. I discussed fifteen works in these sections, and took six of them (all published after the most recent wave of feminism in the United States) as a sample: of the total of 329 works cited in these six sources, there were *zero citations* of works that incorporate sex and/or gender analysis. (The six sources are: McDowell 1979; Nagle 1986; Searle 1984, 1992, 1993; Wiggins 1976.)

[52] This conclusion was urged on me by Jonathan Sills and James Lennox independently.

social settings that are already considered central in most current analytic metaphysics, epistemology, and philosophy of science.

Somehow, though, a unified front is implicitly presented against feminist epistemologists: "objectivity" is of utmost clarity and importance to everyone except the feminists, who are caricatured as disregarding it in order to further their political agendas.[53] This is absurd. Current metaphysics, epistemology, and philosophy of science take no such view of "objectivity" as granted. Feminist epistemologists are challenging themselves and other philosophers to clarify and defend specific concerns about objectivity, truth, and knowledge. Given the problems raised in this paper – that is, the anachronistic ontology of the *ontological tyranny*, the vulnerability of even those who are taken as paradigmatic defenders – Williams and Peirce – of an alternative model that could legitimately resist the relevance of feminist categories and challenges, and the abundant and explicit recognition of certain aspects of "the social" by authors who rarely or never cite or address feminist concerns – the burden of proof is clearly on those who wish to reject the centrality and relevance of sex and gender to our most fundamental philosophical work on knowledge and reality.

[53] E.g., Searle, on the pernicious influence of ideology in feminist scholarship and teaching: "... if you think that the purpose of teaching the history of the past is to achieve social and political transformation of the present, then the traditional canons of historical scholarship – the canons of objectivity, evidence, close attention to the facts, and above all, truth – can sometimes seem an unnecessary and oppressive regime that stands in the way of achieving more important social objectives" (1993, pp. 70–71). That is, political activists don't let facts stand in the way of social reform. It is worth noting that Searle does not cite a single feminist work in this entire piece, even though his examples are almost exclusively about the dangers of feminism in the academy.

11

Science and Anti-Science

Objectivity and Its Real Enemies

INTRODUCTION

As the political activist Ti-Grace Atkinson wrote in 1970: "whenever the enemy keeps lobbing bombs into some area you consider unrelated to your defense, it's always worth investigating" (1974, p. 131).

Such an investigation is the primary aim of this essay. There are several interrelated pronouncements that materialize with mystifying but strict regularity whenever "feminism" and "science" are used in the same breath. These include: feminists judge scientific results according to ideological standards instead of truth and evidence, and are recommending that others do the same; feminists are all "relativists" about knowledge, hence they don't understand or don't accept the basic presuppositions of scientific inquiry; feminists – like many historians, sociologists, and anthropologists of science – wish to replace explanations of scientific success that are based on following the methods of science, with explanations purely in terms of power struggles, dominance, and oppression, and to ignore the role of evidence about the real world; in sum, feminists don't believe in truth, they reject "objectivity" as being oppressive, they are hostile to the goals and ideals of scientific inquiry, and they renounce the very idea of rationality itself.

Given that there is a well-known body of feminist work which has systematically articulated *the negation* of each of the above beliefs in black and white, there is certainly a mystery here. In the past four years, a number of scientists and scholars in the studies of science (historians and

Lloyd, E. A. (1996). Science and Anti-Science: Objectivity and Its Real Enemies. In L. Hankinson Nelson and J. Nelson (Eds.), *Feminism, Science, and the Philosophy of Science* (pp. 217–259). Dordrecht: Kluwer Academic Publishers.

philosophers), have sounded an alarm about the *anti-scientific* nature of feminist scientists and feminist analyses of science, making highly visible accusations against feminists of the sort just enumerated. Had these critics engaged feminist researchers *other* than those who have made explicit their pro-science commitments, their claims might be interpreted as merely poorly-informed, but this is not the case; the authors they claim to engage are among the most overtly pro-science feminists.

My ultimate goal is to identify and examine central assumptions and loci of concern that play important roles in attempts to discredit feminist contributions to the sciences and to science studies. Because feminists share many goals and analytic tools with other social studies of science, I have found that it is imperative to examine how the critics set up their objections to contemporary science studies. My investigation therefore begins with a display of selected quotations attacking science studies in general (Section 1). In Section 2, I suggest a framework for analyzing these attacks, which involves appreciating the inherent tensions among openness about information, the maintenance of social authority and stability, and scientific and democratic ideals; I think these issues lie at the heart of the conflicts and failures of communication between the critics and the social views of science they attack. In Section 3, I analyze several widespread complaints about social and historical studies of science, complaints that our critics raise against feminist approaches as well. I offer diagnoses of the argumentative strategies used, in order to document an important pattern; my goal is to develop conceptual tools adequate to comprehending the origins of the critics' astonishing misrepresentations of feminist science studies.

I review some key features of the relevant set of feminist approaches to the sciences in Section 4, and examine several of the standard attacks on them in Section 5. I first consider some objections that counterpose scientific and scholarly objectivity and the search for truth with feminists' overtly political goals, and argue that those posing such objections are sorely mistaken about both their simplicity and force. I then illustrate and address claims that certain feminist analyses of science are deeply hostile to the aims and standards of scientific inquiry, and that they appeal ultimately to a rejection of reason itself; I document the distortions necessary to sustain such criticisms. I conclude, in Section 6, that the efforts to demonize and dismiss feminist analyses of and contributions to the sciences, amount to illegitimate and unfounded attempts *to exclude participation by fully-qualified colleagues in these aspects of intellectual life solely on the basis of their feminist politics.* Given this dismal conclusion, I must say that

I hold no great expectations of convincing the authors I address that such exclusion is a bad idea; rather, I wish to take the opportunity provided by these critics to clarify certain feminist claims regarding the sciences, and to develop the distinctions and frameworks necessary for the rest of us to move on.

1. THE ATTACK

I begin with a selection of quotations in which crucial aspects of science studies are challenged.

Gerald Holton: citing the agenda of the Nobel Conference XXV (October 1989), from the invitation to participants, which presents the issue of whether: "science, as a unified, universal, objective endeavor, is over. . . . We have begun to think of science as a more subjective and relativistic project, operating out of and under the influence of social ideologies and attitudes – Marxism and feminism, for example. This leads to grave epistemological concerns" (Holton 1993, p. 143).

Holton's book is about an "anti-science movement," which he describes as: "a number of different groups which from their various perspectives oppose what they conceive of as the hegemony of science-as-done-today in our culture. . . . What they . . . have in common is that each, in its own way, *advocates nothing less than the end of science as we know it.*" He defines the goal of these groups as: "the delegitimation of (conventional) science in its widest sense: a *delegitimation* which extends to science's ontological and epistemological claims." Holton lists the four "most prominent portions of the current counter-constituency, this cohort of delegitimators"; at the "intellectually least serious" extreme, he describes the feminist view: "A fourth group, again very different, is a radical wing of the movement represented by such writers as Sandra Harding, who claims that physics today 'is a poor model [even] for physics itself'." For her, *science now has the fatal flaw of "androcentrism"*; that, together with faith in the progressiveness of scientific rationality, has brought us to the point where, she writes, "a more radical intellectual, moral, social, and political revolution [is called for] than the founders of modern Western cultures could have imagined." One of her like-minded colleagues goes even further, into the fantasy that science is the projection of Oedipal obsessions with such notions as force, energy, power, or conflict

(Holton 1993, pp. 152–154, my emphasis). Finally, Holton explains that he has been motivated to write his book because of the social dangers of this "anti-science" movement: "alternative sciences or parasciences by themselves may be harmless enough except as one of the opiates of the masses, but . . . when they are *incorporated into political movements they can become a time bomb waiting to explode*" (Holton 1993, p. 181, my emphasis).

Lewis Wolpert: describing the irrationality that he identifies with *relativism:* "For philosophers of science, and for some sociologists . . . the nature of science and the validity of scientific knowledge are central problems . . . and *some have even come to doubt whether science is, after all, a special and privileged form of knowledge – 'privileged' in that it provides the most reliable means of understanding how the world works*" (Wolpert 1992, p. 101, my emphasis). In his concern to defend and promote scientific thinking, Wolpert worries about the effects of social studies of science: "emphasis on social processes determining the acceptance of a theory . . . can lead one to a relativistic view of science. For if there really is *no rational way of choosing* between rival theories, for choosing between one paradigm or theory and another, then it seems that *science may be a mere social construct* and that a choice of scientific theories becomes like fashion, a matter of taste" (Wolpert 1992, p. 103, my emphasis). In rebuttal, Wolpert insists: "It is not easy to see how the discovery of messenger RNA or the structure and role of DNA could merely be social constructs. They could only appear to be so to someone *ignorant of the complex science involve*" (Wolpert 1992, p. 115, my emphasis). "It is misleading to think . . . that *science is really nothing but rhetoric, persuasion and the pursuit of power*" (Wolpert 1992, p. 117, my emphasis).

M. F. Perutz: ridiculing what he describes as "the line laid down by certain social theorists who assert that *scientific results are relative and subjective,* because scientists interpret empirical facts in the light of their political and religious beliefs, and under the influence of wider social and cultural pressures. They allege that instead of admitting their preconceptions, scientists misrepresent their findings as absolute truths in order to establish their power" (Perutz 1995, p. 54, my emphasis). Perutz also claims that "because references to right and wrong would imply the *existence of objective truth,* they have been eliminated from the vocabulary" of many sociologists of science (Perutz 1995, p. 54, my emphasis).

Paul Gross and Norman Levitt: in their book criticizing numerous approaches to understanding the sciences, they concern themselves with those they label the "academic left," that is, "those people whose doctrinal idiosyncrasies sustain the misreadings of science, its methods, and its conceptual foundations" (Gross and Levitt 1994, p. 9). They are worried about what they see as an *"open hostility toward the actual content of scientific knowledge and toward the assumption, which one might have supposed universal among educated people, that scientific knowledge is reasonably reliable and rests on a sound methodology"* (Gross and Levitt 1994, p. 2, my emphasis). Gross and Levitt claim that this hostility is expressed through adopting "philosophical relativism" and "strong cultural constructivism" about science, which they define as believing that: "science is a highly elaborated set of conventions brought forth by one particular culture (our own) in the circumstances of one particular historical period; *thus it is not*, as the standard view would have it, *a body of knowledge and testable conjecture concerning the 'real' world* . . . [instead,] Scientific questions are decided and scientific controversies resolved in accord with the ideology that controls the society wherein the science is done. *Social and political interests dictate scientific 'answers'*" (Gross and Levitt 1994, pp. 45–46).[1] The central problem with this approach to science is that it *"leaves no ground whatsoever for distinguishing reliable knowledge from superstition"* (Gross and Levitt 1994, p. 45, my emphasis).[2] They explicitly tie this extreme view with feminist approaches to science, asserting that: "Cultural constructivism – in its strong form – is one of the starting points and chief ideological mainstays of the feminist critique of science" (Gross and Levitt 1994, p. 47). Gross and Levitt conclude that the views of science they consider constitute a dire social threat: "What is threatened is the capability of the larger culture, which embraces the mass media as well as the more serious processes of education, to interact fruitfully with the sciences, to draw insight from scientific advances, and, above all, to evaluate science intelligently" (Gross and Levitt 1994, p. 4).[3] In fact, they identify the attitudes of their target social scientific and feminist authors as an "intellectual debility afflicting the contemporary university"; this is extremely dangerous, even genocidal, because the health of our

[1] That is, strong cultural constructivists "view science as a wholly social product, a mere set of *conventions* generated by social practice" (Gross and Levitt 1994, p. 11, their emphasis).

[2] Or, it "affords *no special leverage* among competing versions of the story of the world" (Gross and Levitt 1994, p. 38; my emphasis).

[3] Cf. p. 15, on the potential for these authors having a "great and pernicious social effect."

universities is "incalculably important for the future of our descendants and, indeed, of our species" (Gross and Levitt 1994, p. 7).

Themes

There are numerous concerns raised in the above passages; before proceeding, I would like to identify the key issues I shall address in this paper. Most generally, I propose a *way of thinking about the perceived conflicts* between these scientists and those who analyze the sciences within more inclusive social frameworks. In the following section, I suggest and develop a sympathetic interpretation which addresses the *social and political concerns* that are visible in the opening quotations. Having acknowledged that vital socio-political interests are served by maintaining and enhancing the authority of a scientific worldview, I develop an analysis of the tensions and conflicts regarding the control and dissemination of *information* concerning scientific activities – conflicts that arise inevitably when scientific, sociopolitical, and democratic interests coincide.

The concept of "relativism" seems to be a central obsession of the critics of science studies I quoted above (for brevity, "the critics"). Nevertheless, I will not address the strengths and weaknesses of relativism, *per se;* instead, I shall investigate what these critics believe to be *involved in* a commitment to "relativism" on the part of researchers who examine scientific investigations as social processes. I argue in Section 3 that the *threat* of "relativism," as seen by the critics, derives in part from attributing – in error – a simplistic and false dichotomy to science studies research: namely, that social investigations and explanations of scientific processes and products *necessarily exclude and are incompatible with* investigations and explanations that are presented in terms of standards of evidence, theories, testing, and acceptance which compose the framework of "internal," scientific evaluations of knowledge. I consider several specific criticisms which make use of this fundamental dichotomy, concluding with a discussion of the attractions and pitfalls of this particular response to the "demystification" of scientific activities.

2. INFORMATION AND CONTROL

I will not argue that no reason or no argument could be given that might justify a ban on the pursuit or dissemination of social studies of the sciences. In fact, I take it that such an argument could be reconstructed from the chaotic reactions of the alarmists quoted in Section 1. One version could go like this: since the collapse of the ruling and controlling power of religious organizations in Western society,[4] the authority of *Science* has taken its place, and for good reasons. Human progress rests on increased knowledge of – not superstition about – the natural world, and how to control it. Accordingly, the awe once inspired by religious ritual is now inspired by feats of insight into and control over nature itself. Science now serves the controlling, inspirational, and authoritative roles formerly belonging to Church-states and their military force.

One consequence of the sciences' central social and political role, as described above, is that a certain *image of science* must be maintained in order to ensure social and economic order: Science must be as believable and trustworthy an authority as possible. Furthermore, some special, authoritative, even mythic stature is psychologically necessary to the maintenance of these essential social roles and functions, and that stature is supported by instantiations of the archetypal story line known as the "origins myth." In describing a particular picture of science as an origins myth, my aim is *not* to claim that the events recounted did not occur, but rather, to emphasize the cultural and psychological *functions* of the account in question. The focus is on the *role* of a particular picture of science in providing and maintaining specific social goods, such as consensus or agreement (e.g., about public education), peace or nonconflict, and an active and effective democracy.

Let us assume that scientific knowledge and the people who produce it – scientists, technicians, equipment-designers, and so on – play such a central role in maintaining vital social, political, cultural, and economic stability, safety, and prosperity, that, therefore, protecting their social *authority* is in all of our best interest. It could therefore appear that feminists and radical critics, when they expose and detail the scientific weaknesses and self-serving interests – whether of class, religious, heterosexual, sex, or racial privilege, usually unconscious – of some scientific approaches and some scientists, are actually destabilizing a great deal: the authority of science and scientists has assumed such a pivotal role in the social order that it can be assaulted only at great peril to the safety and well-being of all.

[4] Perhaps precipitated by Darwinism, Freudianism, the atheism of the dialectical materialism of Marx and Engels, or by the rise of industrial capitalism and the urgent need to replace the masses' preoccupation with the afterlife with a preoccupation for worldly goods that would provide the expanding markets essential to capitalism, and so on.

Let us also accept this social and political perspective on the necessity of maintaining a strong, authoritative role for the sciences in public understandings. I think that such a context creates a tremendous tension between the two primary constructive roles of *studies* of the sciences. On the one hand, descriptive accounts of the sciences must serve as the factual basis supporting the myth-level origins stories; on the other hand, studies of science may play vital and detailed roles in the ongoing cultivation and refinement of scientific knowledge and methods.

In essence, one can accept all this and still believe that there is more to scientific investigation and knowledge than a private conversation between Nature and the careful (or bold) and right-minded scientists. Reason and experience tell us that there are good theories, with good supporting evidence, stalled in their tracks for decades because of a favored established theory and its connections to other sciences (e.g., plate tectonics in geology). There are also good theories, with good supporting evidence, rejected for decades because of conflicts – imagined to be irreconcilable – with newer experimental evidence (e.g., the rejection of natural selection as a primary cause of evolution because of breeding experiments). These episodes in the histories of geology and biology, respectively, can be used to teach a variety of lessons, including: true theories don't always look true when we get new evidence; false theories will eventually be rejected in favor of truer ones; being right can pit a scientist against all other established scientists, for life; it's always good to search for more evidence and different kinds of evidence, and so on. I take it that all of these conclusions can be supported by the historical records, even though they may, individually, exemplify conflicting or incompatible lessons about the local workings of scientific inquiry.[5] Hence, there is a great deal of latitude in interest and emphasis when choosing which, if any, of these lessons to spell out using the histories of the sciences. Different lessons are appropriate for different purposes, and the aims of legitimating and maintaining authority can *differ* from the goals of understanding the details of how, when, and why a scientific approach, program, or theory was pursued and then came to be accepted as part of the scientific edifice.

The contemporary situation involving social studies and critiques of science might be better understood by a comparison to military intelligence. A nation interested in commanding global respect for its military

[5] See David Hull's (1988) important analysis and documentation of a variety of dynamics in scientific inquiry.

might will, quite properly, recount and advertise its successes, the superior training of its personnel, and the efficiency, accuracy, and power of its weapons and war machines. A military commander involved in waging war, however, will find detailed accounts of previous battles – with all of the miscalculations, mistakes, losses, and body counts – to be *essential* to success in actually fighting the war.

To illustrate: suppose that we are currently at war, and suppose further that we have already amassed detailed information that can be analyzed to determine which strategies are most effective against our current enemy, and where our own weaknesses lie. We could, quite reasonably, defend a two-pronged policy regarding this information. Specifically, we might deem it important to disseminate the information and to develop careful analyses of it, for those who can use it to further the goals of our side, that is, the commanders who make strategic and tactical decisions. At the same time, we might adopt a policy of general secrecy regarding that information and especially our analyses of it, because we gain a strategic advantage through keeping the enemy ignorant of the things we know about them and their operations, and about our own vulnerabilities. In essence, this is the basis of both the value and the secrecy of military intelligence.

Returning to the sciences, we are now in a position to sketch a potential danger of social studies and critiques of science. There are two distinct sets of potential consumers or audiences of such studies and critiques: practicing scientists and their supporters, who already assign substantial authority to the sciences; and everybody else – people who are not directly involved in the development of scientific knowledge but who must, for the sake of order in society, accept the authority of science. Although research investigating social aspects of scientific activity – including the foibles, weaknesses, or dead-ends of scientific research – may be extremely valuable to practicing scientists and their supporters, it's also true that that *same* research may be used, in the broader culture, to *undermine the authority* of the sciences themselves.[6]

Consider the differences between an historical account celebrating a scientific discovery – in which the brilliance and integrity of the discovers, as well as their ultimate scientific triumph, are emphasized – and one recounting the details of the ego battles, missteps, logical leaps, and

[6] A particularly slick appropriation of such research can be found in the creationist arguments of law professor Phillip Johnson, who is fond of twisting Stephen Jay Gould's evolutionist, in-house criticisms of parts of evolutionary theory to discredit evolutionary biology as a science. Karl Popper's philosophy of science is also cunningly misrepresented and recruited for the creationist cause.

massaging of data *involved* in that discovery. These two historical accounts will inevitably be very different in emphasis, and they may even draw different morals or lessons; nevertheless, because they operate on different levels, with the celebratory account emphasizing the long run, ultimate success, and the equally dramatic detailed one emphasizing the bumps and turns of the journey, they can both be accurate. The *primary* difference is that they serve different purposes, and are aimed at different audiences.

Once we focus on the variety of purposes that might be served by developing and disseminating descriptions and analyses of scientific discovery, struggle, and change, a deep misapprehension on the part of the opponents of science studies can be better understood. In brief, they tend not to distinguish between "demystifying" science and "discrediting" it. The publication of "inside" information about the workings, decision making, and standard assumptions of particular scientific projects, people, or fields, *does make* them vulnerable; specifically, such information *could* be used to discredit these projects, people, and fields, and to challenge any social and political authority they might have. There is nothing in the information itself, however, which necessitates its use to discredit; what the information does do is *demystify* – it opens for examination and understanding the workings of social institutions that look quite mysterious from the outside, just as biochemical information and theories opened the living cell. Demystification is the shared purpose of all scientific investigation, including investigations into the operations of scientific investigation. Indeed, this is why science studies are important; if we are to be able to *compare* various systems and methods of investigating nature and producing scientific knowledge, we must understand them.[7] Even the critics we're considering here wish to discriminate among various instances of scientific processes, rejecting Nazi science, denouncing racist science, and science used toward the sole end of individual profit.[8] Nevertheless, they consistently attack others for attempting to "demystify" the sciences; this puts the critics in the untenable position of defending scientific authority

[7] In describing the value of social analyses of science to scientific success, Sandra Harding writes: "we can hold that certain social conditions make it possible for humans to produce more reliable explanations of patterns in nature just as other social conditions make it more difficult to do so," (1992, p. 7). Cf. Gross and Levitt's claim: "scientists welcome the sort of 'social' explanation that examines minutely and honestly the intellectual, attitudinal, and . . . the moral preconditions of culture that encourage and sustain the practice of science" (1994, p. 128).

[8] E.g., Holton (1993, pp. 114–123, 155–156, 181–184); Gross and Levitt (1994, p. 110); Wolpert (1992, Ch. 8).

by insisting on its "mystification."[9] We need to understand the conflicting interests and motivations which have led to this unstable position.

Internal Tensions

A more incisive analysis of the various uses of information would help us to see who is using inside information to further the sciences' objectives and authority, and who is using it with the genuine intention – or result – of undermining that essential authority. The critics' most profound error is that they have misidentified feminist and social scientists as being utterly *hostile* to science, rather than accepting them as allies who genuinely share their concerns and foundational political goals regarding science.[10] Those who advance the sciences are in a uniquely difficult position because of a double pressure – from scientific and democratic ideals – for openness and participation; this is why it is so damaging when their reactions are insufficiently refined, and in some cases border on paranoia – seeing enemies where there are pro-scientific allies.

In summary so far: historical and methodological accounts of science have long served in the roles of legitimating origins-myths, and this is fine, because such legitimation is necessary to public confidence and social order. These accounts can also serve a vital role in recruiting talented individuals to engage in scientific enterprises, and they are also naturally going to be the focus of those responsible for maintaining and funding scientific institutions. But studies of science can also be undertaken to provide more complete and accurate pictures of how the sciences actually work, how they succeed, how they fail, and what the most conducive conditions and standards are for attaining any scientific goals – this necessarily involves a "demystification" of science – as a science of science, it demystifies the workings of an otherwise magical-looking system.

An inevitable product of the investigations into science is, thus, *inside information* which, like military intelligence, has the capacity to be used in various ways, toward various social and political ends. *It has this capacity precisely because the sciences are so intimately tied to political authority.*

[9] For example, Gross and Levitt claim that the aim of the "oppositional social critic" is "to demystify science and topple it from its position of reliability and objectivity" (1994, p. 50; see also p. 234). Later, they write of "the emerging dogmas of the left concerning the innate fallibility of Western science" (1994, p. 243). Surely, Gross and Levitt do not want to defend a position wherein the sciences are *infallible*; they elsewhere claim that success in science requires "unremitting self-scrutiny and attention to the possibility of error" (1994, p. 27).

[10] See Gross and Levitt's section heading, "The Face of the Enemy" (1994, p. 34).

Everyone involved in debates about the nature, authority, and proper roles of the sciences understands this on some level – the creationists, social activists such as feminists and environmentalists, and the critics I discuss in this paper. It is precisely because of the fundamental, architectural role of the sciences in political, economic, and social life that social activists *target* the social authority of specific scientific programs and specific scientists.[11] No one in the United States who heard any discussion of *The Bell Curve* should doubt that contemporary science is as powerful a political player as any in action today. What remains at issue, then, is how this inside information should be analyzed, toward what ends it should be put to use, and how far and in what manner it should be disseminated.

There is thus an inevitable set of tensions facing scientists and those of us who wish to support the authority of the sciences versus, for example, the authority of particular religious principles. The standards of openness in science, public education, and a democratic political system, all make it difficult for scientists to maintain control over information regarding the actual doing of science, and over anyone who chooses to analyze that information. The model of military intelligence – analysis for insiders, pep talks for the troops, and secrecy and disinformation for the enemy – is therefore not one of the available options for the channeling of appropriate information to the right places.

Generally speaking, there are a variety of goals involved in human communication, and a central one of these goals is certainly to disseminate truth rather than falsehood; additional goals include cultivating broad rather than narrow perspectives, focusing on long-term or distant consequences as well as immediate ones, and updating and maintaining broadly shared vocabularies for the purposes of business, civic life, and social stability. The critics who object to feminist and social analyses of the sciences are sympathetically read as focusing on the *potential harm* of such studies to social life, economic prosperity, and the civic responsibility which is essential to democratic government. It isn't at all clear how studying sciences in action will generate this harm, but the assumed mechanism seems to be something like the following: social studies, especially feminist studies, of the sciences sometimes challenge the accuracy and scope of specific scientific conclusions and inferences; by doing so, they're undermining the *authority* of scientific knowledge in the wider culture.

[11] For instance, Barbara Ehrenreich, Stephen Jay Gould, Donna Haraway, Ruth Hubbard, Evelyn Fox Keller, and Richard Lewontin, to mention a few prominent American scientist-activists from the natural sciences.

Hold it right there. How and why do challenges to specific instances or aspects of scientific activity amount to undermining the political and social authority of scientific knowledge itself? Doesn't that depend on the basis or standards which are used to generate the challenges? In other words, if appeals to empirical evidence, consistency, and other scientific standards are the substance of such challenges, then they are properly seen as operating *within* the sciences, and as such, legitimating them. The undermining and delegitimation of the authority of science is genuinely challenged, in contrast, when the *basis* of a challenge is, for example, a fundamentalist religious view, such as the belief that evolutionism teaches atheism.[12] Again, it is a mistake to equate demystification, and any potential improvements in the activities of science that may come to mind with that demystification, with delegitimation and the rejection of scientific authority *überhaupt*.

This vital distinction between ameliorative reform and all-out rejection is being blurred by the critics' reactivity and defensiveness.[13] Holton, for example, advises, unselfconsciously: "for specific facets of anti-science, see the essays . . . in . . . *Counter-Movements in the Sciences*" (Nowotny and Rose 1979), apparently not noticing that the book is about cases *within* the sciences. Gross and Levitt threaten to eject humanities and social studies from the university unless they become more hospitable to being judged by natural scientists (1994, pp. 245–257). Regaining control over *how the sciences are understood* is the aim of Gross and Levitt's entire book, and it is an echo, in less sophisticated wrappings, of Holton's and Wolpert's conclusions, as well. So here is the tactic: mobilize pro-science constituencies to oppose the academic and sociopolitical legitimacy of those who demystify science. But this can't be right; *some* analysis of inside information is desirable for the effective functioning of the sciences themselves. In fact, Wolpert makes this very point, while not seeming to notice the resemblance between his own list of important issues to investigate, and the issues prominently addressed by feminists.[14]

[12] E.g., a popular bumper sticker in San Diego: "Creation: God said it, I believe it, and that settles it."

[13] "The central appeal of [science studies] is the pretext it provides to *disparage the natural sciences* – to dismiss their astounding achievements as so much legerdemain on the part of a ruling elite" (Gross and Levitt 1994, p. 240; my emphasis).

[14] "What is required [of science studies] is an analysis of, for example, what institutional structures most favour scientific advance, what determines choice of science as a career, how science should best be funded, how interdisciplinary studies can be encouraged" (Wolpert 1992, p. 122). See section 4.

So what is the plan? Is it that informed and intelligent analyses and criticism of sciences should remain behind closed doors? Not tolerated in the academy? Does such criticism paradoxically *encourage* superstition – even when one of its aims is to uncover or reveal unsupported beliefs and assumptions within the sciences which might be called superstitions themselves? Given the overriding value of openness of inquiry and information to both a scientific and a democratic worldview, there ought to be very good reasons – and no viable alternative – for limiting such openness.

The recent attacks on science studies have come out openly with their political and social worries, which must be read as their reasons for wanting to limit openness about inside information on the sciences. Their appeal to the importance of the authority of science has turned out to be very valuable, because they have simultaneously revealed their lack of insight into the overall situation involving friendly, democratic, and hostile uses of information. Furthermore, it makes inaccessible the option of taking the "high road": other people are contaminating discussions of the sciences with their political concerns, whereas real science is not appropriately dragged into the political mud.[15] If anyone is in a position to adopt this approach, it is *not* the critics we're discussing, because of the explicitness of the social and political goals of their own writings. Still, understanding the internal tensions inherent to this situation can help us to remain sensitive to the complications and possibilities for misunderstanding that will be inherent when two parties have differing perceptions of the security, dominance, and legitimacy of any particular sphere in which the exercise of scientific authority plays a role with substantive social consequences.

I conclude this section by sketching a few strategies, borrowed from the context of intelligence work, for managing "inside" information to reinforce the authority of the sciences.

Strategies

Control: The most direct way to keep potentially damaging information from being used against the sciences is to keep it secret or private, or, even better, not to gather or analyze it at all.

[15] "Modern science is seen, by virtually all of its critics, to be both a *powerful instrument of the reigning order* and an ideological guarantor of its *legitimacy*" (Gross and Levitt 1994, p. 12; my emphasis). Do they think the sciences play important, legitimating, social roles, or not?

Discredit: Discrediting supposedly-inside information is also very effective. This can be done through: (1) claiming that it is not accurate; or (2) claiming that the purveyor is unqualified and incompetent; or preferably, both.

Exclude: The essence of exclusion is to define a potential source of information or analysis as *hostile* – as an enemy – and thereby to justify their exclusion from any legitimate access to or discussions of that information.

Each one of these strategies has its costs, because each one is potentially in conflict with the ideals and interests of both scientific inquiry and a democratic political system. Thomas Jefferson, quoted approvingly by Wolpert, makes clear the appropriate democratic solution: "I know no safe depository of the ultimate powers of the society but the people themselves, and if we think them not enlightened enough to exercise that control with a wholesome discretion, the remedy is not to take it from them, but to inform their discretion" (qtd. in Wolpert 1992, p. 170).

In the following section, I will discuss how critics of social studies of science impose a condition of exclusivity on social and "internal" scientific explanations of science, even though, *prima facie,* such explanations ought to be considered complementary. I think this move is usefully seen as a setup for a struggle for control over information, especially when it is combined with a perception that any demystification and inside understanding of the workings of science are necessarily damaging to scientific authority and functioning. The problem is that, according to scientific standards themselves, a certain amount of internal analysis is necessary and desirable. As Wolpert concludes, "the Greek commitment to free and critical discussion was essential for science to flourish . . . once one rejects understanding and chooses dogma and ignorance, not only science but democracy itself is threatened" (Wolpert 1992, p. 178). Moreover, to read all demystification as the "enemy use of information" is to make a grave tactical error: it goes against scientists' own principles, and is hence hypocritical; it also underestimates the power of the sciences to withstand scrutiny and to absorb and reform any picture of the sciences' workings.

3. REASON VERSUS RELATIVISM

3.1. The Exclusivity Doctrine

One of the best ways to understand these critics is to examine their attachment to a certain dichotomy, a division of intellectual labor, in which social explanations and "internal" ones are seen as exclusive and competing – rather than complementary – explanations. The basic move is to ascribe to authors in science studies a commitment to a *unitary* and *exclusive* model of explaining events in scientific practice: *all* such events are to be explained completely and exclusively in social terms, allowing no complementary and interlaced explanation in terms of the scientists' own ("internal") beliefs, theories, and evidence.

Let me tell a story. Imagine that we set out to study a diversity of religious rituals practiced in a variety of human societies. We can go in and watch and describe what people are doing, what they won't do, what they respect, what they say, what they refuse to do, and so on. And we also may investigate by asking, by inference, by history, by interview, by any number of ways – what these people are doing and what it means – that is, *why* they're doing what they're doing. In other words, we need to understand both their various beliefs, *and* their beliefs about the relation between their beliefs and the world, in order to fully comprehend what they're doing, in order to make sense of – give an account of – these events, these religious rituals. Now, I take it that others would agree that these observations and investigations of these ritual behaviors are recognizable as part of a general area of study or approaches called Anthropology.[16]

It has been recognized as an important step in understanding the vast varieties of human ways of living, that we understand what human actions *mean* to the person doing them. Also, grasping such meanings must include the theories and assumptions about the ways the world works and the way the world is. Such detailed study of the comparison of actual beliefs and belief-systems – for example, in religious rituals – is probably most analytically and competently handled by theologians, who are familiar with the many aspects of comparing religious belief systems. Still, these studies of the belief systems are only part of the story; they are a necessary part, but only a part.

I think that there is an *exact* parallel to the practice of science, and if we're interested in as full and as complete an understanding of science as possible, I believe that (just as in the comparative religion case) some sorts of social and anthropological investigations are necessary, and that without them we do not have a full and complete account, on *anybody's* standards. This also means that the theoretical background and assumptions, and the beliefs about the relations between the theory, the

[16] Or "cultural anthropology," "comparative religion," and so on; it makes no difference to this discussion what we call these endeavors.

theoretical backgrounds, and the aspects of the world being investigated, are important. Any claim that theoretical problems and their objects are essential factors in an account or explanation of scientific events sounds right to me, in exact parallel to the religious case. I take it, though, that the objection we're considering is that social or anthropological studies of scientists *neglect* these central factors – the beliefs about theory and object, and the truth, accuracy, and evidence for these beliefs – in concentrating exclusively on the social, psychological, economic, or political contexts of scientific work. Potentially, *all* social analyses of scientific activity suffer from this fatal flaw.

The real question is whether the pursuit of social or anthropological descriptions of scientific activities and ways of life is *incompatible* with discussing the content of sets of beliefs and theories, and the reasons and standards that are used to evaluate them. An examination of the *contents* of beliefs would give us only an incomplete understanding of spiritual practices; for the same reasons, a summary of scientific beliefs would give us a very incomplete understanding of scientific practices. As far as scientific theories and their objects go, natural philosophers have spent centuries in detailed investigations of the theoretical problems, the objects, their relations to a set of scientific beliefs and systems, and so on. It's not as if nobody's looked at those problems. That's precisely where the action has been.[17] The concern seems to be, rather, that the social descriptions are intended as *exclusive*. Are social and anthropological accounts of science supposed to give a *complete account of science?* If such completeness is not being attributed, why would Gross and Levitt complain that: "To concentrate on the idea of empirical science as a manifestation of cultural and political imperatives is to *omit* important dimensions of the story, both human and philosophical" (Gross and Levitt 1994, p. 68, my emphasis)? Similarly for Wolpert's objection that a "scientific discovery cannot be judged only in social terms but must also take into account the new understanding or knowledge it provides" (Wolpert 1992, pp. 113–114); using a social approach to scientific knowledge, he complains, "*says nothing* about the belief's contribution to understanding, its correspondence with reality or its internal logical consistency" (Wolpert 1992, p. 110, my emphasis). The crucial difference here is between "saying nothing" and saying *there is nothing else to say*.

It thus seems possible that the critics do not understand the incomplete nature of the internally focused explanations they favor. Analyses

[17] In other words, the natural philosophers have served as the theologians here.

of scientific theories and evidence and explanations are not made inaccessible or irrelevant by the existence of anthropological or sociological approaches; rather, they are complementary. Philosopher of science Sandra Harding makes the point very clearly: "Nature causes scientific hypotheses to gain good empirical confirmation, but so, too, does the 'fit' of these hypotheses' problematics, concepts and interpretations with prevailing cultural interests and values. It is a *maximally objective understanding* of science's location in the contemporary social order that is the goal here. This is *far from a call for relativism.* Instead it is a call for the *maximization of criticism of superstition, custom and received belief* for which the critical, skeptical attitude of science is supposed to be an important instrument" (Harding 1992, p. 19, my emphasis). Without a persuasive reason why social studies cannot contribute or add completeness to our available reasoned understandings, there is no argument – especially not one based on incompleteness – against pursuing them. Because any relevance or intellectual value of the critics' discussions rests on their providing *reasons* for rejecting the intellectual legitimacy of specific approaches to describing the sciences and scientific activities, the rest of this essay will be concerned with developing clues and hints regarding such reasons and arguments, and evaluating the results.[18]

In sum, an implausible and unpromising explanatory doctrine, in which social and scientific/evidential explanations are seen as strictly mutually exclusive and individually complete, is being used by the critics to cause much of the trouble here. (For the sake of brevity, I shall hereafter refer to this explanatory doctrine as "the exclusivity doctrine.") I shall suggest that many of the reactions against science studies, especially the accusations tying "relativism" to "irrationality," are best understood as arising from the exclusivity doctrine. The most important worries revolve around the vital roles of *truth* and of the *reasons* that scientists have for holding specific beliefs and theoretical commitments. The fundamental concerns seem to be that social analyses and critiques of science *neglect* the *content and truth of the scientific theories* they discuss: because content, truth, and evidence constitute the very foundations of science, any analysis which neglects them must fail to reveal anything significant about the sciences

[18] Again, these authors *claim* they're not against pursuing the questions asked in history, philosophy, anthropology, or sociology of science; they are only against how these studies are actually done (Gross and Levitt, 1994, p. 69; but see nn. 7 and 9). Their proposals for a proper or more appropriate standard of practice for these studies will be discussed in later sections; at this point, however, the burden of proof is on them to reject the present standards in science studies.

themselves. I argue that it is the critics of science studies who, paradoxically, tend to *insist* upon the exclusive nature of social and evidential explanations, rather than the authors to whom they attribute this position.[19] The paradox is illuminated by highlighting the strategic value of attributing the exclusivity doctrine; it can be an effective basis for both the "discredit" and "exclude" strategies introduced in Section 2.

3.2. Illustrations: The Burdens of Proof

In this section, I interpret and analyze two standard objections to the "relativist" and "strong cultural constructivist" views, as they are perceived by our critics of science studies. Although the objections differ in their details and targets, they share an underlying theme: they *insist* that social explanations and "internal," evidential explanations of scientific activity are mutually exclusive – in spite of indications by the authors they cite that they reject the exclusivity doctrine.[20]

The Control of Description

Gross and Levitt attack philosopher and anthropologist of science Bruno Latour by accusing him of holding the view that "laboratory politics *accounts for science as such* and is the *real story* behind the emergence of scientific theories" (Gross and Levitt 1994, p. 58; my emphasis); "Latour's reports on the activities of scientists *are* to be accorded factual status," they exclaim, "while scientists' reports on nature are *not*" (Gross and Levitt 1994, p. 58; their emphasis).

One predictable objection to pursuing the social study and critique of scientific practices is to challenge the supposed "privileged viewpoint"

[19] Although it is possible to locate passages in which social explanations do seem to be presented as complete (or as replacement) explanations, even those science researchers most closely linked to this thesis in the early 1980s have since published views that make no such claim. James Griesemer has pointed out that, construed within their context, the more sweeping claims are most obviously read as efforts to carve out disciplinary turf; Wolpert's casual defense of the necessary excesses of discipline-establishment is therefore of great interest: "Of course there was some resistance to the new ideas and the molecular biologists were evangelical in trying to persuade others. They undoubtedly also used rhetoric" (1992, p. 104; James R. Griesemer, comments, public lecture by Paul Gross, November 1995, UC Davis).

[20] One point to note immediately is that the critics' insistence on the exclusivity doctrine raises the stakes for both of the supposed completely social and completely "internal" or evidential explanations; in fact, it virtually demands an overextension of explanatory tools to areas in which they are inappropriate.

of such research. "Why should we believe," the usual argument goes, "so-and-so's claim that socioeconomic factors are playing a significant role in the scientific community pursuing or believing theory or research program *A* rather than *B*?"

This is an important and interesting challenge. But note what the objection relies on setting up as the alternative: "why should we believe so-and-so's claim" . . . *instead of the scientists' own claims* regarding why they're pursuing *A* rather than *B*. There are several serious problems here. Most obviously, it is a mistake, as I argued above, to *assume* that social explanations are posed – by Latour, in this case – as *replacements* for all the reasons adopted by any individual scientists. Second, the implicit alternative – that the scientists involved can simply reflect on and report their reasons for pursuing *A* rather than *B* – does seem to assume that scientists have super-human powers or insight. As philosopher of science Sandra Harding has pointed out, if "the 'science of the natural sciences' is best created by natural scientists . . . [then] the sciences would be the only human acitivity where science recommended that the 'indigenous peoples' should be given the final word about what constitutes a maximally adequate causal explanation of their lives and works" (1992, p. 14).[21]

It is important to remember that we need make no assumption that *others* necessarily have privileged information about scientists' motives, any more than we need grant the exclusivity doctrine. Note that the appeal of Gross and Levitt's objections, quoted above, relies utterly on attributing the exclusivity doctrine to Latour, even though he doesn't hold it.[22] This is a distraction from the fundamental issue, which is simply that giving *unquestioned and exclusive authority* to scientists' own descriptions of their actions and motives is – given common-sense acquaintance with the complexities of human self-understanding – an impoverished and silly way to understand scientific activities. The *relevant standard* here comes from social and anthropological studies *of other complex human activities*.[23]

[21] Harding also emphasizes the unsuitability of the training of natural scientists for the task at hand: "Natural scientists are trained in context stripping, while the science of science, like other social sciences, requires training in context seeking" (1992, p. 16).

[22] They conveniently deny the fact that *things* – parts of the world that scientists interact with during their investigations – play essential roles in Latour's 'networks' (Gross and Levitt 1994, p. 59).

[23] Remember that we are considering the *legitimate scope* of pursuing social and/or anthropological analyses of the processes of scientific research and explanation. The decisive problem with the question concerning the "privileging" of scientists' own explanations is that it changes the subject; *it is irrelevant*. The specific challenge we're considering asks, in its strictest form, whether *any* anthropological or social explanation of human activity

Once this is explicit, it is difficult to see how these critics could defend a blanket intellectual rejection of such studies. Should models of market forces in economics be applied to grains, gold, women, antibiotics, and petroleum but never to scientists and their grant money? Should admittedly ethnocentric studies and interpretations of social structures, values, and cultural practices be performed on non-European groups of people but never on groups of scientists? Should correlations between biological sex, wealth, and labor be investigated and detailed for communities around the world but never for scientific laboratories? The weaknesses – predictive, descriptive, and explanatory – of these ways of studying our social lives are well known, and I agree that we would do well to maintain the visibility of those weaknesses. The opponents of social studies of the sciences, however, cannot *legitimately exempt* the sciences *as subjects of study,* without some additional argument.

The Control of Judgement

In highlighting another example of their insistence on the "total opposition" between social and internal scientific explanations of scientific practices I appeal to certain passages quoted by or paraphrased by Gross and Levitt; even though these passages are offered as evidence that Steven Shapin and Simon Schaffer, in their *Leviathan and the Air-Pump,* claim to give an exclusive and complete explanation for the science they investigate, they support no such claim. Rather, they appear to map out claims of *partial* explanation.

For example, Gross and Levitt quote Shapin and Schaffer's own claim: that "these political considerations *were constituents* of the evaluation of rival natural philosophical programmes" (1985, p. 283, qtd. in Gross and Levitt 1994, p. 63; my emphasis). Their paraphrases of Shapin and Schaffer's conclusions include the following: "the nascent Royal Society was, from the first, the creature and deputy of a political and social viewpoint"

ought to be *believed.* The problems are legion: the epistemic nightmare of attributing intentions (and to what?); the metaphysical issues concerning the relative reality of subatomic particles, organic chemicals, organisms, and their psychological states (and the relations among these); and the ontological status of groups and their parts, systems and their components, functions and roles and goals. These are, indeed, some of the central issues in the philosophy of social sciences. But the actual question being addressed has been lost in this methodological swamp; it is, as posed by its enemies: are anthropological studies and social critiques of science legitimate topics of university study? (See Chapter 9 of Gross and Levitt 1994, in which they threaten humanists and social scientists with expulsion from universities because the physical scientists won't tolerate them anymore.)

(Gross and Levitt 1994, p. 64); "[Robert Boyle's and his coworkers'] supposedly empirical rules, it is said, constituted a specific social practice" (Gross and, Levitt 1994, p. 63);[24] and, finally, "The [Royal] society's supposedly objective science is thus to be read, *in large part*, as a construction of its ideological commitments" (Gross and Levitt 1994, p. 64, my emphasis).

Note that claims to empirical adequacy, predictive power, or scientific explanatoriness of a scientific theory or approach are *nowhere here reduced to or exclusive of its* political commitments. That is, the position which Gross and Levitt attribute to Shapin and Schaffer *here* does not, in fact, include the exclusivity doctrine; rather, they attribute a more moderate view, in which both "objective scientific" and "ideological" explanations play roles.

In order to support their claim that Shapin and Schaffer adopt the exclusivity doctrine, Gross and Levitt must ignore not simply the immediate context, but the entire book surrounding the statement they chose to represent its most damning conclusion. Gross and Levitt claim that the authors think that this case study embodies a radical relativism; they write that the following passage "is Shapin and Schaffer's last word on the general epistemic principle that their particular case is supposed to illustrate." Quoting Shapin and Schaffer: "As we come to recognize the conventional and artificial status of our forms of knowing, we put ourselves in a position to realize that it is ourselves and not reality that is responsible for what we know" (1985, p. 344, qtd. in Gross and Levitt 1994, p. 65). Gross and Levitt then remark, "So, in the end, we come back to the dichotomy – fallacious in that it posits total opposition between 'reality' and 'convention' where there is, in fact, intense and continuing interaction – so favored by Latour and other constructivists" (1994, p. 65).

I shall say more about this attribution of the exclusivity doctrine to the "constructivists" in a moment. First I must admit, however, that of all the sentences in Shapin and Schaffer's 440-page book, this one almost begs to be plucked out and misinterpreted. Nevertheless, there are compelling reasons against construing it to mean: *"it is exclusively ourselves and not reality at all* that is responsible for what we know," as Gross and Levitt apparently do. One serious problem with such an interpretation is that it fails utterly to take account of the ongoing discussions and debates within the history of science of which this book is a part: in the wake of centuries of historical accounts of science in which the overpowering

[24] Surely Gross and Levitt would not want to deny this.

majesty of nature and the beauty and perfection of its rational order were portrayed as the engine driving scientific progress itself, recent historians have been developing more accurate and complete accounts, focusing on the specific decisions, activities, and desires of the investigators, and on the social and cultural contexts in which their interactions with nature came to yield scientific knowledge. There is, thus, a corrective force to Shapin and Schaffer's book regarding the history of science itself. More importantly, there is a plausible and reasonable interpretation – supported by abundant textual evidence – of the quoted passage, as follows.

In the context of the case study in the book – the development and use of the "air pump," or vacuum container – "artificial" signifies that we – members of a scientific community – crafted our apparatus (specifically, the air pump) through which we experiment and come to know things, whereas "conventional" signifies that the procedures and experiments that were worked out by Boyle had to be argued about and agreed on through social processes of persuasion and interpretation – otherwise, the demonstrations wouldn't be understood *as* experiments. That this is a reasonable understanding of the situation is clear from Shapin and Schaffer's examination of the explicit debates about those methods, apparati, procedures, and interpretations; they document the arguments and the winning moves. In this context, it does *not* amount to any radical rejection of scientific standards, truth, or the existence of the real world, to conclude, as they did, that what we know *depends on* the existence and acceptance of specific equipment and procedures – that is, their existence and acceptance is *necessary to* our knowledge; hence *we,* and *not exclusively the reality independent of us,* are "responsible" for what we know. The point is both simple and incontestible: more than the *existence* of reality is necessary for us to know it; it has been there all along, and yet we have not known it until very recently; in order for us to know it, it must necessarily come within the purview of whatever *methods and experimental apparatus we do have,* which are, themselves, *our own creations.*

I find this to be a fairly typical example of Gross and Levitt insisting on the exclusivity doctrine. Interestingly, Gross and Levitt are aware that their criticisms appear designed to discredit and exclude, and they make an effort to display their good faith and openness regarding science studies by actually engaging Shapin and Schaffer on one issue: explaining why Thomas Hobbes's views were not accepted.

This little argument is extremely revealing because Gross and Levitt display their own ideas about the standards by which social studies of science should be judged. In brief, Gross and Levitt argue that Hobbes's repeated failure in mathematical arguments "provides a concrete and substantive reason, *in contrast to an ideological one,* for Hobbes's notoriety in scientific circles" (1994, p. 67; their emphasis).[25] Note that they have introduced a dichotomy between scientific or mathematical explanations, and social or "ideological" ones. It is clear that *they think* the two types of explanation are exclusive; if, they say, Shapin and Schaffer had addressed Hobbes's failure as a mathematician, they "would have put themselves in the position of conceding the existence of sound, objective reasons for deciding at least some scientific controversies" (Gross and Levitt 1994, p. 67). But Shapin and Schaffer did not deny the existence of any reasons that might be categorized today as defensible, scientific ones, nor is there a reason given to think that social explanations cannot be sound or objective. After Gross and Levitt legislate that *no social explanation is compatible with a [potential] 'internal' one,* they go on to condemn Shapin and Schaffer for "insisting that all such disputes are ideological" (1994, p. 68).

One could interpret all of these exchanges as a consequence of a benign lack of subtlety combined with defensiveness about maintaining the myth-serving aspect of the history of science. That would be a mistake, as Gross and Levitt appear to be very sensitive to exactly what is wrong with both a myth-preserving view and the view they foist on the "constructivists"; remember, the target of much of Gross and Levitt's criticism is the dichotomy that "posits *total opposition between 'reality' and 'convention' where there is, in fact, intense and continuing interaction"* (1994, p. 65; my emphasis). This vision of "intense and continuing interaction" between reality and convention is the basis of the very texts they've attacked. Although Gross and Levitt claim to reject the completeness and exclusivity of social and other explanations, their own arguments and conclusions rely extensively, and sometimes completely, upon just such exclusivity; furthermore, they seem so wed to their beliefs about the "constructivists" that they miss all of the textual evidence supporting a more moderate reading, even when it is given in their own paraphrasing (e.g., previously quoted uses of "constituents of" and "in part"). The climax comes near the end of their book, where they admit that "those who insist that science is driven by culture and by politics . . . are not for that reason

[25] One hotly debated question at the time concerned the proper *role* of mathematics, scientifically; the issue was especially pressing, given the prominent place that Descartes had given mathematics in the definition of knowledge itself, contrasted with the deficiencies of his physics. Gross and Levitt ignore this.

alone to be dismissed as wrongheaded. On the contrary, these assertions, if 'driven' is replaced by 'influenced,' come near to being truisms" (Gross and Levitt 1994, p. 234).

In sum, Gross and Levitt utilize the full battery of strategies against those they have branded as anti-science. In the text just considered, they misrepresent the social history of science they wish to reject; specifically, they ignore even *their own* references to partial causes, influences, and implications of shared social and empirical accounts. Such a pattern shows something far more interesting than its evident scholarly impairment: it clearly embodies the strategies of (1) attributing programmatic hostility; (2) discrediting interlocutors for being ignorant of the sciences; and (3) attempting to gain control over the terms of "legitimate" social studies of science. This last is never so clearly revealed as when Gross and Levitt propose their own standards and limits for studies of science: "inasmuch as the specific content of the [cultural-constructivist] thesis challenges the reliability of scientific conclusions . . . and inasmuch as it does so, roughly speaking, on the basis of the same argumentative paradigm as scientists use in practice, the logic, evidence, and pertinacity of the thesis must be weighed against that of specific scientific arguments" (Gross and Levitt 1994, p. 49). Gross and Levitt demand that, in order to make their case, "cultural constructivists must demonstrate that their arguments for unreliability outweigh those of conventional scientific papers for reliability *in the realm of phenomena addressed by the latter* [their emphasis]. They must show that their arguments are stronger than those put forth by Professor X in his paper on the role of transforming growth factor beta in the morphogenesis of the optic tectum, while simultaneously outweighing those of Dr. Y in his monograph on the classification of compact Lie group actions on real projective varieties! If they are to demonstrate that their arguments *contra* science are anything but sheer bluff, then clearly *they must play on the scientists' court.*" (Gross and Levitt 1994, p. 49; my emphasis).[26] And Gross and Levitt are confident that such arguments will fail, because, "to put the matter brutally, science *works*." (1994, p. 49; their emphasis). Hence, in spite of their lip-service to "intense and continuing interaction," Gross and Levitt don't grasp or don't accept the essentially complementary nature of social, cultural, and evidential explanations.

[26] Given this view of appropriate explanation and evidence, we must wonder about the ingenuousness of Gross and Levitt's disclaimer that "working scientists are not entitled to special immunity from the scrutiny of social science" (1994, p. 42).

Wolpert has similar difficulties, even though his is generally a much more subtle and penetrating understanding of scientific investigation than Gross and Levitt's. Although Wolpert notes that Andrew Pickering's study of particle physics is self-consciously oriented toward the social aspects of scientific activities, Wolpert concludes, with some astonishment, "there is really nothing in his analysis that reflects such an approach or that is in conflict with an image of scientific advance *that scientists themselves would readily find acceptable*" (1992, p. 116; my emphasis). Thus, he seems to be suggesting that real constructivist analyses of science would, necessarily, *conflict* with scientists' self-understanding. This situation is illuminated by Wolpert's description of Pickering's account: "[it] shows just the sort of complex interactions between theoreticians and experimentalists that *one might expect.* . . . What Pickering does make clear is the symbiotic relationship between theoreticians and experimentalists: both are looking for new opportunities to advance their work" (1992, p. 116; my emphasis). Here, Wolpert is clearly speaking *as an insider;* these are insider's expectations, and an insider's understanding of the "symbiotic relationship" and the goals of advancing one's own work. In other words, because Pickering has "merely" made public "inside information," in an accurate fashion that is acceptable to (some) insiders, Wolpert concludes that he has not, in fact, taken a genuinely "constructivist" approach; the obvious but abandoned alternative is that social and internal explanations of scientific exploration and change are *not* mutually exclusive or inherently conflicting.

Again, the overall tension is between a view of science used to inspire, promote, and legitimate, and a framework which can be used to inform scientists, philosophers of science, and others interested in the intimate details of science-in-action; a deep conflict can arise only out of insistence on the legitimacy of only one of these approaches to science.

3.3. Strategic Intelligence

The distinctions I've illustrated can be used to reveal a profound problem with the basic position which the critics wish to maintain. Consider the pictures of science that these authors wish to defend and disseminate: they explicitly claim that their goals for such public communications are social, economic, and political, just as they criticize their chosen opponents for *undermining* the accomplishment of these same goals. So far, so good; differences in strategies for accomplishing public, shared goals are the bread and butter of civic and policy debates, and this is their proper forum in democratic societies. *But these authors, in entering these debates, deny that they are even participating in them.* This surrealist move is accomplished by claiming that their views are the *only* views that are rational and true: they are not debating policy, they are announcing fact, and anyone disagreeing with them is necessarily advocating falsehood and perpetuating ignorance.[27]

If Wolpert, Perutz, Gross, and Levitt are making factual claims about the social, conceptual, and historical facets of the sciences, then the intellectual and scientific standards they champion demand that the truth or accuracy of factual claims must be evaluated with respect to evidence, as must accusations regarding the falsehood of the factual claims of their opponents. Instead, the critics tend to emphasize that their targets neglect scientific and evidential accounts of the sciences – we have seen that this complaint is damaging only in the context of the exclusivity doctrine, which they impose on the authors they criticize.

Through assuming and imposing an "exclusive and complete" picture of social versus "internal" evidential explanations, the scope of social studies of science becomes indefensibly narrowed. Under such an explanatory doctrine, the legitimate topics of social studies of science encompass only whatever the scientists chose to regard as *unscientific*, that is, as *not explainable* using the internal standards of that science – in which case they might better be called "studies of non-science." Fundamentally, this move amounts to setting the detailed decision-making, exploratory, and evaluative activities of scientists off-limits as subjects of analysis, which in turn eliminates the possibility of legitimately pursuing many social critiques of science.[28] Because any overt demand that scientists have a special, exempt status with regard to studies of human social systems and cultures is recognized as being in extremely bad form – as anti-scientific itself – the critics I examine offer disclaimers to the effect that no exemption is being demanded, while they simultaneously implement the strategies of secrecy, discrediting, and exclusion, all of which are justified by an appeal to the security of the democratic and social order.

[27] "We are accusing a powerful faction in modern academic life of intellectual dereliction. This accusation has nothing to do with political correctness or 'subversion'; it has to do, rather, with the craft of scholarship" (Gross and Levitt 1994, p. 239).

[28] See Gross and Levitt's appeal to scientists' right to judge, as experts, *all* work concerning "scientific methodology, history of science, or the very legitimacy of science" (1994, p. 255).

Hence, the critics engage several of the available strategies for coping with the tensions regarding scientific and sociopolitical control of information are engaged in these criticisms; I find that their use has serious disadvantages, not the least of which is their dubious strategic effectiveness. In the contemporary U.S. context, it is unclear whether insisting on the purity and apolitical sanctity of the sciences is a viable intellectual, political, or social option. What Gross, Wolpert, Perutz, and others, seem to be in denial about is that the cat's already out of the bag. Just as the rash of political assassinations, the Vietnam War, the Pentagon Papers, Watergate, Nixon's resignation, and changing journalistic standards contributed to a long-term, if not irreversible, public mistrust in elected officials and public institutions, so the spectacles of corruption and waste in the manufacture and design of our best military technology, lack of responsiveness within all parts of the health and medical technology professions during the first decade of the AIDS pandemic, and lying and cheating for money and prestige within the top universities and research institutions in the world, and even the waffling on diet guidelines regarding cholesterol, eggs, or oat bran – all have produced a public mistrust of both the disinterestedness and competence of scientists in general, and thereby of science itself.[29]

These events suggest that there might be an aspect of *scapegoating* to the fury that Holton and others direct toward anyone they perceive as "demystifying" science. The scientists involved in the incidents mentioned above needed no science studies sleuths to appear foolish, greedy, dishonest, biased, overreaching, or blinded by ambition; they did it all by themselves, and often in public view. The contributions of the science studies researchers whom Gross et al. target might best be described *not* as "revealing" and advertising the inside information that such unseemly

[29] The majority of adult Americans receive their information about the world from TV news, with radio news running second. Among the science scandals aired on national network news within the past twenty-four months, I would mention: the manufacture of data for the Pittsburgh Breast Cancer study; the reinstatements of eggs into the recommended anti-cholesterol diet; the well-publicized omission of women from nearly all of the most extensive and expensive heart disease studies, which led to a special initiative by Congress; and earlier, the Dalkon Shield devastation; the fanciful claim by President Ronald Reagan that there is no evidence that radiation causes cancer (see Philip Fradkin 1989); or the revelations of the horror of radiation experiments done on unsuspecting civilians from the 1940s through the 1970s. The public perceptions of some of the scientists involved in these events is far from the genius with special insight into nature, and closer to Drs. Frankenstein or Mengele.

events have happened, but rather as investigating the possibilities that such things are built into the social systems of the sciences as they stand: they might be structural, institutional, and predictable. In that case, they would not be freak accidents, any more than Oliver North was a rogue elephant single-handedly violating U.S. policy.[30]

Conclusions

Scientific activities are unique ways of arriving at knowledge which depend on free exchange of information and criticism; they also *present unique problems,* precisely because of this necessity of openness of information and responsiveness.

I began this paper with an acknowledgment that there are legitimate and important issues that arise from the tensions between the essential social and political roles of the sciences and the standards internal to democracy and to science. Scientists and supporters of science must therefore reckon with the fact that we don't have control over all information about scientific activities which might be relevant to various parties, including genuinely hostile ones; moreover, this information *cannot* be controlled the way that military intelligence is controlled, because of the open critical exchange essential to the sciences themselves. Specifically, scientific standards make it difficult for scientific leaders, or those most interested in maintaining scientific authority, to use the strategies of propaganda. We could read the desire to "restore the old dominant view" (Perutz 1995, p. 54), as a *wish* that they could somehow shield the general public (or the university student) from these portraits of science at work, because *none of the usual mechanisms are available to make sure that propaganda goes out and secrets stay home.* They cannot – because of a fundamental, overriding interest in open scientific communication – classify very much as "top secret"; hence, they can't control access to the nitty-gritty details of scientific processes, successes, or failures. In other words, the first strategy listed in Section 2, "control," is out; the burden therefore falls to heavy use of the strategies of discrediting and excluding. These are, I shall argue in the rest of the paper, the primary maneuvers that the critics have used in their reactions to the feminist authors they cite.

[30] The fact that "executive deniability" has been an essential part of CIA operations policy since its inception is well documented; see John Ranelagh's (1986) sympathetic history, *The Agency: The Rise and Decline of the CIA.*

4. FEMINIST CONTRIBUTIONS TO SCIENCE AND SCIENCE STUDIES

4.1. *Objectivity, Truth, and the Standards of Scientific Communities*

With the foregoing analysis in hand, it becomes much easier to see that it is *not* a reasonable or adequate response to feminist contributors to science, to point out that they are politically motivated. Playing fair – that is, according to standards for scientific conduct – a *scientific* alternative, challenge, criticism, or commentary, must be evaluated and answered scientifically. There has been an enormous amount of confusion about this seemingly obvious point, but for now, you don't have to be a radical philosopher of science to see this; it is sufficient to buy John Herschel's (1831) distinction between the context of justification and the context of discovery.[31] Under this quite conservative view of science, the *source* of an alternative hypothesis or a criticism is seen as irrelevant to its scientific merit. This standard and ideal of scientific practice is important for two reasons. First, it highlights the fact that any refusal to consider and respond to feminist scientific contributions embodies a double standard: chemists wondering about the structure of hydrocarbons did *not* dismiss Kekule's benzene-ring structure because it came to him in a dream of a snake swallowing its own tail (1965, p. 9); indeed, tolerance of a wide variety of explicitly political ideologies (among male scientists, anyway) has been one of the hallmarks and points of pride of the international communities of twentieth-century scientists, and rightly so, because it fulfills a *standard of open-mindedness* essential to the practice of science itself. Second – and this is a closely related point – those who must respond to feminist ideas in the sciences do not have to accept any feminist views regarding the sources of the issues, problems, or omissions being debated: that is, they don't have to believe any feminist doctrines whatsoever, in order to *address* feminist scientists. Furthermore, they don't have to agree with feminist views to agree with some feminist scientific claims and conclusions; one would expect, in fact, to have many results of feminists doing science accepted *as* simply "good science."[32]

[31] I must note that this distinction has come under sustained criticism within philosophy of science, especially by feminists. My focus here, however, is on the most conservative views of science held by working scientists. The point is that even under these views, objections to the feminist *source* of specific scientific contributions violates the canons of scientific conduct.

[32] See Sections 4.2 and 5.2 for elaboration. For the most recent work on why sexist science is not properly characterized as "bad" science, see *Synthese*, 104 (September 1995).

A quick glance at developmental geneticist and feminist Anne Fausto-Sterling's aims and methods makes it abundantly clear that she is *not* recommending that "political interests dictate scientific answers," either in the science she criticizes or in that which she promotes. In presenting her analyses of scientific claims regarding human sex differences, Fausto-Sterling advises her readers to apply perfectly ordinary scientific standards: "look at the data, think about the logic of the argument, figure out how the starting questions were framed, and consider alternate interpretations of the data" (1985, p. 10). Fausto-Sterling is signaling precisely that she is interested in enforcing adherence to usual standards of "good science": in her criticisms of specific claims regarding sex differences, she objects to their "gross procedural errors," their "striking errors in logic," and their "inaccurate understanding of biology's role in human development" (1985, p. 8, 60). Similarly, neurobiologist and feminist Ruth Bleier argues that otherwise-good scientists "have shown serious suspensions of critical judgement in interpretations of their own and others' data," that they have ignored the known "complexity and malleability of human developments" and that they have made "unsubstantiated conjectures," not one of which "is known to be descriptive of scientifically verifiable reality as we know it today" (1986, p. 149; Bleier, 1984).

Feminist contributions to and critiques of the sciences have also long been concerned with the structure and dynamics of the self-corrective processes of producing scientific knowledge: the now-standard feminist argument has been that it makes for *better science,* to encourage the training and full participation of informed researchers with a variety of background experiences, preconceptions, and viewpoints, precisely because such inclusion will encourage a wider variety of working hypotheses, as well as more thorough challenge and testing of any given scientific hypothesis or theory that is under consideration.[33] Philosopher of science and feminist Helen Longino, for example, has argued that "scientific method involves equally centrally the subjection of hypotheses and background assumptions to varieties of conceptual criticism and the subjection of data to varieties of evidential criticism" (1993, p. 266). In her explication and endorsement of standards internal to the community of science that lead to its objectivity, Longino writes: "Effective criticism produces change,

[33] See esp. Longino (1993), pp. 257–272; Longino (1990); Dupre (1993); Harding (1991); Harding (1993), pp. 49–82; Harding (1995); Longino (1995); Nelson (1990); Nelson (1993), pp. 121–159; Tuana (1995); Tuana (Ed.) (1989); Wylie (1988); Cf. Feyerabend (1975); Mill (1859).

and a community's practice of inquiry is *objective* to the extent that it facilitates such transformative criticism" (1993, p. 266; my emphasis).

Thinking about "objectivity" in scientific knowledge is very difficult due to the multiplicity of meanings and contexts of the term itself; I have delineated four basic meanings that are in wide current use, as follows.[34] When applied to knowers, "objective" means *detached,* disinterested, unbiased, impersonal, or invested in no particular point of view; in such cases, objectivity is not a property of whatever is known through these methods. Other uses of "objectivity" are more complicated, in that they involve relations between things: when "objective" means *public,* publicly available, observable, or accessible (at least in principle), some relation between reality and knowers is involved. Similarly, when "objective" means *existing independently* or separately from us, it directs us toward some relation between us as knowers and the reality we're trying to gain knowledge of. Finally, there is a current meaning of "objective" as really existing, "Really Real," or the way things really are. This last usage is supposed to apply no matter what the relations are between reality and knowers. I have found that untangling the various meanings of "objectivity" is absolutely essential to interpreting and evaluating the numerous claims involving feminism, objectivity, and knowledge.

Returning to Longino's discussion of the attainment of objectivity in scientific thought, she emphasizes the mechanics of self-correction in her descriptions of how individual variation in scientific opinion "is dampened through critical interactions whose aim is to eliminate the idiosyncratic and transform individual opinion and belief into reliable knowledge" (1993, p. 265; See also Longino 1990, esp. Chs. 4 and 9). Hence, her emphasis is on a certain sort of detachment, and on public critical agreement. Longino lists four key features necessary to the knowledge-productive capacity of scientific communities: "avenues for the expression and dissemination of criticism; uptake of, or response to, criticism; public standards by reference to which theories, etc. are assessed; and equality of intellectual authority" among qualified practitioners (1995, p. 384).[35] On

[34] Treated at length in Lloyd (1995).

[35] Cf. Wolpert, on science's "rigorous set of unstated norms for acceptable behaviour": "Included in these norms are the ideas that science is public knowledge, freely available to all; that there are no privileged sources of scientific knowledge – ideas in science must be judged on their intrinsic merits; and that scientists should take nothing on trust, in the sense that scientific knowledge should be constantly scrutinized" (1992, p. 88). Like Longino, Wolpert emphasizes the community-level process over the individual traits of researchers: "leaving aside the question of whether scientists are more objective, rational,

this analysis, the objective and self-correcting nature of scientific inquiry, which is counted among its most profound strengths, is actively reinforced by feminist participation.[36]

4.2. Good Science

We need to investigate briefly an interesting dynamic that is at work when feminist contributions to the sciences that *are* compelling and accepted, are written off as "good science at work" and as having nothing whatsoever to do with feminism.[37] It might be thought, especially in light of the discussion in Section 4.1 concerning scientific standards of evaluation and their divorce from the origins of any candidate scientific contribution, that feminist scientists must abandon any claims regarding their own ideological commitments. This may be true with regard to the ultimate evaluation and acceptance of any particular feminist scientific claim; that is, feminist scientists have not been in the business of demanding that their scientific claims be accepted or rejected on purely ideological bases. Quite the contrary, as I reviewed earlier.

One crucial point is easily lost, though, in the context of feminist insistence on applying rigorous scientific standards, and that is the vital role of the participation of feminist thinkers in the celebrated self-corrective processes of science. As Fausto-Sterling argues, in her discussion of feminist corrective contributions to medical and behavioral sciences: "These ideas, although they may represent good science, arose in the context of a vast and multiply branched political–cultural movement, that of modern Western feminism. [To apply a purely] good versus bad science analysis is to ignore the important role feminism has played in forcing the reevaluation of inadequate and often oppressive models of women's health and behaviour" (1985, p. 213; her emphasis). Fausto-Sterling then elaborates on the crucial role feminism has, in fact, played in the dynamics

logical and so forth, scientists have developed a procedure in which there are free discussion, accepted standards of behaviour and a means of ensuring that truth will, in the long run, win. Truth will win in the sense that open discussion and observing nature *constitute the best way of making progress*" (1992, pp. 122–123; my emphasis).

36 The critics' evasion of this core component of feminist views is discussed in Section 5.2.

37 Gross and Levitt assert: "there are as yet no examples . . . of scientific knowledge informed, reformed, enhanced by feminism" (1994, p. 112). Their strategies for dealing with the numerous feminist contributions to the sciences they subsequently cite are instructive: briefly put, if feminist work is persuasive and is accepted as correct, it's simply good science; if not, it's bad science tainted by ideology. In other words, the feminist contributions to science are either *not feminist* or *not contributions*.

of supporting corrective scientific challenges: "In the past, legions of highly trained doctors and scientists have failed to see and criticize what is wrong with the biomedical and behavioral models of female behavior. Why? Because . . . they had no alternate framework within which to develop new sight. Feminism provided that new vision, allowing many scientists – even those who do not consider themselves political feminists – to move in a new direction" (1985, p. 213).

The feminist scientists and philosophers of science we've considered have argued that the self-critical and participatory aspects of scientific inquiry, when combined with empirical evidence and its analyses, are able to produce the *most reliable and objective scientific practices;* furthermore, the full participation of qualified feminist scientists and critics constitutes an essential part of this self-corrective dynamic.

4.3. Illustration

Contrary to the accusation that feminist views of science involve sweeping antiscientific and anti-reason biases, actual feminists' contributions to and critiques of the sciences are projects which address particular assumptions in particular versions of some people's ways of doing science; they frequently argue that, on the standards accepted by the scientists in question, the goals of a particular pursuit in science would be better served by rejecting or changing certain assumptions and practices. Consider the following example of feminist science at work.

Primatology

The changes wrought in scientific understanding of primate behavior through feminist interventions would be difficult to overstate.[38] I focus here on a single article written by statistician and primatologist Jeanne Altmann, and published in 1974. In "Observational Study of Behavior: Sampling Methods," Altmann surveyed sampling methods that could be used by researchers in the field, who typically have little or no control over the movements or conditions of the animals they are studying (1974). For decades, primate studies had relied heavily on sampling *ad libitum,* which tended to result in reports of rare or dramatic events, or in emphases on events of particular interest to the observer; such observational

[38] See Haraway (1989) for a comprehensive bibliography and analysis. I have borrowed from Haraway's discussion of J. Altmann in presenting this case.

results could not, Altmann pointed out, provide the evidence necessary to answering a host of crucial questions, including those involving differences in behavior patterns among individuals and across subgroups, such as male and female, adult and adolescent. Altmann also provided a devastating methodological criticism of a widely used and supposedly sophisticated sampling method, in which the occurrence (or nonoccurrence) of a particular type of event was recorded; one consequence of her statistical critique was an immediate and irrevocable reduction in the scientific import of the work of some of the field's leading researchers.

But Altmann's primary contribution was positive: she articulated the procedures and advantages of a method that she dubbed "focal animal sampling," in which a focal animal or group of animals is followed for a pre-set time period, and all occurrences of a specified action or interaction are recorded. As Donna Haraway points out, "The embarrassing truth was that many of the regularly cited field studies ... both gathered and analyzed data in a way that did not justify the conclusions reached" (1989, p. 307). In essence, Altmann raised the standards of evidence accepted within the entire community of field primatologists, and her paper became one of the most cited in the entire modern literature on animal behavior.[39]

Nothing in the foregoing reveals the fact that Altmann herself was acutely dissatisfied with the skewed visions of primate social-structure that arose out of pervasive biases toward focusing on male animals and on dominance interactions, and that her active participation in the modern U.S. feminist movement contributed to her awareness of the significant theoretical implications of these biases.[40] Nevertheless, her revision of sampling practices cleared the way for pursuing a central problem in available theories of primate evolution; specifically, Altmann thought that differential reproductive success – the "motor" of natural selection – was much more significant among females (and much less so among males) than had ever been acknowledged. As leading primatologist Sarah Blaffer Hrdy described the situation twelve years later: "changes in methodology (e.g., focal animal sampling of all individuals in a group) and the emergence of long-term studies played critical roles in revising male-centered models of primate social organization" (1986, p. 135–136).[41]

[39] Cf. Wolpert, in discussing the scientific importance of a contribution: "As the mathematician David Hilbert once expressed it, the importance of a scientific work can be measured by the number of previous publications it makes it superfluous to read" (1992, p. 86).

[40] As Haraway (1989) documented and analyzed, pp. 304–310; also, personal communication with Jeanne Altmann.

[41] Cf. Fedigan (1982); Strum (1987).

Note also that Altmann's feminist contribution does not fall along the lines that some might expect. Altmann's methods enforce – by any account – *an increase in objectivity and precision,* and she makes no mention of her beliefs regarding the sexism in the prevailing sampling methods she so effectively replaced. This illustrates that feminist contributions cannot be identified by their content; rather, feminist scientists have changed those fields in which they have participated through active engagement with particular experimental, analytical, and theoretical problems.

4.4. Making Enemies

In stark contrast to their portrayal by the critics, feminists contributing to the sciences and science studies insist – repeatedly, consistently, in detailed and appropriate contexts – on the centrality of truth, evidence, and objectivity; furthermore, some have effected an increase in intellectual and evidential standards within their own scientific or philosophic communities.

I would like to address briefly the nature of "the feminist menace"; *why* have these feminists provoked the grotesqueries of misinterpretation which I document in the next section? Given that the feminist scientists and analysts in question operate within, and appeal to, the explicit goals and standards of their own scientific and intellectual disciplines, they cannot be read as introducing conflicting bases on which knowledge claims should be evaluated – in contrast, for example, to the creationists. Some feminist concerns do center around social abuses of scientific authority, but even these do not challenge the fundamental desirability of a scientific basis for knowledge claims, that is, scientific authority itself. The authoritative mantel of science is supposed to be hard-earned; the science involved must be good science, if appeals to its authority are an important part of the sociopolitical landscape.[42] Hence, social abuse of scientific authority arises when: (1) specific scientific theories and conclusions which are serving social roles do not fulfill standards of good science (i.e., the "scientific authority" is not earned); (2) efforts to defend these "scientific" results utilize tactics intended to exclude and control qualified critics, rather than relying on the standards of scientific inquiry

[42] Wolpert makes this very point, when he identifies the sole social responsibilities that scientists must take on: "Scientists have an obligation to make the reliability of their views in . . . sensitive social areas clear to the point of overcautiousness. And the public should, wherever possible, demand the evidence and critically evaluate it" (1992, p. 163).

(i.e., the scientific standing is not defended scientifically); and, finally, (3) the scientific results in question reinforce and legitimate the social status quo (i.e., the science is used socially to protect those in power). Therefore, the feminist criticism of the social abuse of scientific authority can only be seen as threatening to scientific authority by identifying scientific interests with specific social ones.

It is also essential to understand that there are clear remedies to suspected cases of social abuse; the accusation of abuse of authority can be blunted effectively by *responsive* scientific communities, with an emphasis on self-correction, fair-mindedness, and intellectual responsibility. The canonical "Mistakes were made; scientists are human. We've responded rapidly and effectively to well-grounded objections," does involve some demystification and admissions of fallibility, and such "demystification" may reduce, temporarily, the untouchability of scientific authority concerning the matter at hand. The alternatives, however – corruption, dishonesty, and strong-arming – are unlikely to do the authority of science any good in the long run.

Thus, it seems that the threatening aspect of feminist and social analyses of the sciences is the revealing of the interdependence of social and cultural context, scientific practices and products, and the actions of individuals; this is, in fact, clear from the critics' confusion about the exclusivity doctrine. They seem willing to acknowledge that, in *general,* scientists are social, human, beings, but unwilling to admit that the social, cultural context played any significant role in any particular case. In spite of the resulting aggravation, I would argue that this resistance serves a vital function: it is a method of maintaining scientific standards within science. In fact, many of the debates in which feminist scientists have engaged have involved efforts to apply critical scientific standards to scientific claims; in other words, they were ordinary contributions to scientific debates. There is, hence, no necessary "abuse of authority" accusation involved in feminist participation in ongoing scientific debates. The actual abuse of scientific authority arises only when the social-authority functions of science *overrule* the knowledge functions; this is why demonstrations of responsiveness are the most effective response to accusations of abuse.

Feminist contributions to the sciences, then, are best described, not as challenges to the legitimacy of science, or even as accusations of the abuse of scientific authority, but, rather, as developments of the information and analyses necessary for temporary, local, scientific self-correction. The self-corrective capacities of the sciences serve as the *foundations* of scientific legitimacy and authority. Short-term fallibility and the willingness to admit error are the price to be paid for maintaining both individual authority within the sciences, and the social authority of the sciences in a democratic state. Such willingness to admit error is everywhere a matter of judgment, of course, a careful balancing of whether to keep on, give in, or investigate alternatives. For scientists to admit they were incorrect in the past, other members of the scientific community who share investigative interests must provide the evidence, interpretations, and alternatives, in addition to exerting pressure to meet the relevant scientific standards. Let us now consider some responses to feminist efforts to do just this.

5. IDEOLOGY VERSUS TRUTH

5.1. Critics

Margarita Levin: "*The real threat to feminist ideology,* it turns out, is the scientific method itself, with its *promise of objectivity* no matter who the scientist is"; Levin describes "the first of three fundamental errors that form the basis of the feminist account of science," which is that feminists commit the genetic fallacy, that is, they "confuse something with its origin and reject it on that basis." She then outlines the feminists' "second basic error, which is their failure to take seriously the fact that so-called masculine science *works.* Science makes predictions that can be and are verified every day" (Levin 1988, p. 100; my emphasis). What feminists just don't understand, she claims, is that "the *self-correcting character of the scientific method,* with its emphasis on observation, the replication of experiments, and open discussion, insures that ['deviations from the ideal of objectivity'] will eventually be seen as such" (Levin 1988, p. 100; her emphasis). Ostensibly against feminist views of science, Levin explains that masculinist metaphors that might have helped formulate scientific theories are "completely irrelevant to the verifiability and accuracy of scientific theories inspired by those metaphors" (Levin 1988, p. 104; my emphasis).

Clifford Geertz: "The worry is . . . that the *autonomy of science, its freedom, vigor, authority, and effectiveness, will be undermined* by the subjection of it to a moral and political program – the social empowerment of women – *external to its purposes*" (Geertz 1990, p. 19; my emphasis). In considering the long-range prospects for feminist science, Geertz concludes that its development "depends most critically on how the *tension gets resolved between the moral impulses* of feminism, the determination to correct gender-based injustice and secure for women the direction of their lives, *and the knowledge-seeking ones of science,* the no-less-impassioned effort to understand the world as it, free of wishing, 'really is'" (Geertz 1990, p. 23; my emphasis).

Paul Gross and Norman Levitt: "Recent feminist theorizing about the sciences . . . contains heavy doses of dogma," and like "other dogmatisms," it is "beyond the reach of rational argument" (Gross and Levitt 1994, p. 111). Feminists aim for "the dethronement of Western modes of knowledge and their claims to objectivity . . . [and] champion 'women's ways of knowing' . . . [and think] that Western paradigms have been effectively demolished" (Gross and Levitt 1994, p. 38). More specifically, "the relativism of cultural constructivist doctrine is the perfect tool for *discounting science as biased or corrupt if and when it inconveniences one's political program*" (Gross and Levitt 1994, p. 162; my emphasis). "It is a commonplace among relativists of all kinds [including feminists] *to ignore or dismiss the self-correction process by which good science survives and bad science* – that which is not verifiable by others of different tastes and tendencies – *vanishes in due course*" (Gross and Levitt 1994, p. 123; my emphasis).

One persistent theme elaborated by critics of feminist contributions to science, such as those sampled above, is that feminists pursue ideology instead of truth; their activities are, therefore, in conflict with the aims of science. There are two readily available versions of this claim: Feminists formulate and evaluate specific scientific claims by having their ideological commitments *override* standard scientific goals of discovering truth. That is, feminists use ideological criteria *in place of* scientific criteria. A more radical interpretation is that feminist ideology itself involves a *wholesale rejection* of scientific standards such as objectivity, scientific methods, and scientific criteria of evaluation. On this view, feminists are *anti-science,* and are wholly incapable of contributing to or furthering scientific goals or discovering truth.

Another, partly orthogonal, set of claims centers around the theme of science's *self-corrective* capacities and their role in the effectiveness and genuine objectivity of scientific inquiry. One version of this complaint is that feminists simply don't understand that scientific inquiry is self-correcting. A more damaging accusation is that feminists reject or devalue the scientific process of self-correction. Because their critics recognize the essential links between the self-correction of science and its *objectivity*, feminists are seen as either *neglecting or violating* the very core of scientific inquiry – scientific objectivity.

The critics' claims are striking for two reasons: first, their attributions so vividly conflict with what the feminists in question have actually said and done; and second, the misattributions are so clearly aimed at discrediting and excluding these feminists. I am *not* suggesting a conspiracy of any kind; rather, I have argued that the strategies of discrediting and excluding are fairly general ones for dealing with the publication of inside information which is perceived as threatening. In the following subsection, I detail several of the specific forms these strategies take in the hands of our critics, including: discrediting by claiming that feminists place ideological goals before scientific ones; excluding through omitting or obscuring *scientific* challenges offered by feminist scientists; discrediting by using a definition of "scientific objectivity" intended to disqualify feminists; excluding by painting feminist scientific contributions as essentially and radically hostile to scientific aims; and finally, excluding by obscuring the feminist sources of scientific self-correction, that is, "good science." I conclude, in Section 5.3, with some of the most overt attempts at exclusion.

5.2. Strategies: The Discrediting and Exclusion of Feminists

Pursuing Ideology and Pursuing Truth

Feminism might be seen as an endorsement of a specific set of doctrines or dogmas. One fear is that feminists will give top priority to pursuing their political goals – and to protecting the truth of any dogmas that they deem necessary to those political goals – and will reject genuinely open inquiry into the scientific strengths and weaknesses of the dogmas themselves. If they do this, they separate themselves from the goals of scientific inquiry, which puts open investigation into the truth of any and all empirical claims as its top priority. Because feminists cling to specific dogmatic views – involving the eliminability of certain gender roles, the social aspects of the development of sex differences – they disqualify themselves as participants in open, scientific inquiry.

The portrayal of feminists as dogmatic can be found in Gross and Levitt's discussion of Longino's analysis of the interpenetration of "contextual values" (i.e., norms and values of the social and cultural context), and "constitutive values" (norms and values internal to the sciences), within the scientific investigations of hormonal influences in sex differences. Having quoted Longino's claim that the studies in question "are vulnerable to criticism of their data and their observation methodologies," they launch the following complaints:

> nowhere in the body of Longino's work do we find identified specific, recognizable flaws in the data and the methodologies.... Indeed, the criticisms are not directed toward those at all. Instead, they are either banal (e.g., the argument that data from rodents should not be used to infer processes in people), or indictments of the investigators for making value judgements about departures from sex-stereotypical behaviors.... Led to expect serious criticism of data or methodologies, we find, not cooked data, uncontrolled experiments, or statistical gaffes, but implicit attitudes claimed to have been detected – by a hypersensitive anti-essentialist. (Gross and Levitt 1994, pp. 145–146)

At this point, Gross and Levitt deliver their verdict: "By and large, the logic here is that, since the conclusions are unacceptable by feminist lights, the science must be flawed" (1994, p. 146).

Gross and Levitt accuse Longino of rejecting specific scientific results *because* "What Longino is really after is a way of doing science that will negate *any* possibility of biological determination" (1994, p. 147, their emphasis), that is, that her overriding commitment is to a specific scientific conclusion for ideological reasons. Gross and Levitt's conclusion, that "science as-it-is becomes, for such critics, an intolerable constraint, terrible danger" (1994, p. 147), provides the real leverage used to discredit feminist scientific criticisms *as scientific*. The accusation is so common that it has a standard form:[43]

1. feminists reject particular scientific findings and explanations *exclusively because of their political content or implications*;
2. this is a wrongheaded basis on which to evaluate scientific claims;
3. its use demonstrates that the feminists are *not being scientific* – that is, are not using scientific standards – either because they don't know how, or because they are motivated to "twist the facts" or "reject the truth" in order to attain their political ends (which is precisely what they criticize others for doing);
4. therefore, because the feminist contributors are not being scientific, no scientific evaluation of or response to their claims is warranted or merited.

[43] Other instances can be found in: Levin (1988); Ruse (1981); Searle (1993).

A brief diversion into Gross and Levitt's discussion will lead us to the primary point that is being so insistently buried here. They claim to have sought – in vain – for any substantive scientific content in Longino's discussion of studies on hormonal influences on human sex differences; therefore, they were (reluctantly) forced to conclude that the feminists were being unscientific. Under these circumstances, it becomes rather suspicious, if not deceptive, that they ignore Longino's repeated references to the detailed scientific critiques offered by Bleier and by Fausto-Sterling (Longino 1990, pp. 119, 127, 131, 134; Longino and Doell 1983, pp. 206–227).

The fact is that Bleier and Fausto-Sterling offer detailed analyses of "the cooked data, uncontrolled experiments, or statistical gaffes," which Gross and Levitt claimed were nowhere to be found.[44] In fact, Gross and Levitt do cite Fausto-Sterling's relevant work once – calling it "her polemical book" – when they accuse her of firmly denying the existence of "significant biological differences between men and women" (1994, p. 125).[45] Their mischaracterization of Fausto-Sterling's book – and their nonengagement with her detailed scientific objections – play a pivotal role in the soundness of their entire analysis, as does the invisibility of neurobiologist Bleier's work.[46] Evidently, Gross and Levitt feel that they need this genuinely scientific work not to exist; in the presence of undeniably scientific and feminist contributions, their blanket dismissal – as I've outlined it in this section – of any possible relevance of feminist contributions will fail.

The use of the standard "purely ideological rejection of science" accusation is quite risky: even the appearance of possible scientific objections must be eliminated; otherwise, the authors would be expected to engage their colleagues in ordinary scientific debate. This brings us to the

44 (See block quote, earlier.) Bleier (1984); Fausto-Sterling (1985), pp. 133–141. The work of Harvard biologist and feminist Ruth Hubbard is also very important, especially: Hubbard, Henifen, and Fried 1982; Hubbard 1990.

45 Because Fausto-Sterling acknowledges the existence of biological differences between males and females throughout her book, it remains mysterious how Gross and Levitt could defend this statement, unless they put all the weight for its truth on whatever they mean by "significant."

46 In spite of Bleier's high scientific status as a research scientist in the specialty in question at a major research institution (the University of Wisconsin, Madison), the book in which her scientific objections to this research are summarized and defended, *Science and Gender* (1985), is not listed in Gross and Levitt's bibliography. Nor do they ever mention, in parallel to their treatment of Ruth Doell (Longino's early coauthor and a biologist), and Fausto-Sterling (a senior research scientist at Brown), that Bleier is even a scientist.

fundamental point: feminists have everything to gain from being included in such "ordinary" scientific debate. This has been, in fact, a primary feminist goal. And Gross and Levitt have – inadvertently – shown some awareness of the importance of this goal to feminists, in their determination to resist its satisfaction. Far from advocating an abandonment of the principles of open, critical inquiry, feminist scientists have consistently attempted to enforce them; far from being "anti-scientific," respected scientists have aimed to improve their sciences through demonstrating and insisting upon the highest standards of scientific evidence and justification.

Contrary to the caricature of dogmatism discussed earlier, feminist commitments to open-mindedness and the value of fair and responsible scientific inquiry have served as the linchpins for feminist contributions to the sciences.

Feminism as Anti-Objectivity

As reviewed in Section 4.1, appeals to the strengths of *objective methods,* and to the centrality of *objectivity* in our searches for knowledge about reality play pivotal roles in feminist scientific self-understanding. The accusation of dogmatism – with its lack of fidelity to genuinely open and objective inquiry – is sometimes used within the sciences to *disqualify* feminist participants from full membership in the scientific community, in the following way:

1. Our scientific projects are concerned with objective knowledge and objective reality; hence, by definition, they must involve objective inquiry;
2. feminists are promoting specific ideologies, or social values; this is what it *means* to be a feminist;
3. but being "objective" just means being *free* of values or biases, or commitments or ideologies; therefore:
4. the pursuit of objective scientific inquiry is *incompatible* with simultaneously pursuing a specific ideology or any particular social values.

(In brief: the type of science we are pursuing is objective. Ideology is not. Therefore it is impossible to pursue both at the same time; they're incompatible.)

The most important thing to notice about the standard argument I outlined above is that it utilizes only part of one of the meanings of "objectivity," which has currency today. If any of the other meanings listed in Section 4.1 are substituted into the argument, it becomes invalid. Consider a plausible alternate premise: (3*) Being "objective" just means pursuing truths about things that exist completely independently from us.[47] *It simply doesn't follow* that "(4) The pursuit of objective scientific inquiry is *incompatible* with simultaneously pursuing a specific ideology or social values." In order to sustain this conclusion, additional arguments must be provided that pursuing truths about independently-existing parts of the universe is logically, psychologically, or at least statistically incompatible with performing behaviors aimed at achieving a particular social organization or dynamic (i.e., acting as a feminist). One of the challenges facing this further claim is that feminists have already provided arguments and examples in which using feminist analytic tools has *contributed* to the long-term empirical adequacy and objectivity of scientific research (see Section 4.3). Consequently, those who wish to exclude feminist analyses *a priori,* because of some apparent conflict between feminism and "objective" scientific method, ought not base their arguments on simplistic, misleading, and inaccurate views about the nature of "objectivity."

Feminism as Anti-Scientific

Gross and Levitt have claimed that "strong cultural constructivism," in which science is viewed as "a wholly social product, a mere set of *conventions* generated by social practice" (1994, p. 11; their emphasis) is a "chief ideological mainstay" of feminist views of science (1994, p. 47). Remember, Gross and Levitt say that on such an approach: "Scientific questions are decided and scientific controversies resolved in accord with the ideology that controls the society wherein the science is done. Social and political interests dictate scientific 'answers'" (1994, p. 46).

I have already noted that Gross and Levitt's position has required them to neglect the plain scientific content in the feminist views they dismiss. Before it can be taken seriously, any characterization of feminist contributors to science which postulates some deeply anti-scientific agenda, which is accompanied by rejection of the most basic scientific standards, will have to reckon with the scientific competence and scientific attitudes displayed by those feminists.

[47] Gross and Levitt appeal to *this* meaning of "objective" when discussing "the objective nature of the phenomena" of physics, while depending on the "detached" view when criticizing Keller (1994, pp. 128, 141). Wolpert, on the other hand, rejects the "detached" view as desirable for science (1992, p. 92).

Feminists versus Self-Correction: Good Science at Work

In a bizarre twist in the plot, the very dynamic of participation and self-correction emphasized by the feminists we've been considering has been used *to dismiss* the significance of feminist participation in the sciences;[48] the issues raised deserve scrutiny.

One claim is that feminist scientists just don't comprehend that the processes, methods, and standards of scientific inquiry make it self-correcting. This accusation is uninteresting because it is patently false, although some might find it amusing that Gross and Levitt, in their untouchable faith that feminists are enemies of science, apparently believe themselves to have trapped in a contradiction a group of feminist scientists who appeal to the importance of "controlling for [gender] bias" in order to improve biological science (1994, p. 122). Gross and Levitt gleefully lecture: "to 'control for bias' is an ancient house rule of empirical science.... It is one of the hallmarks of the 'good science' that the postmodernist critics of science [among whom Gross and Levitt include even those feminists who claim to be empiricist] disparage" (1994, p. 274).[49]

But we could imagine a more substantive claim along the following lines: feminists wish to have a free rein to scrutinize the sciences, but wish themselves to be exempted from the critical scrutiny which enables science's self-correction. In fact, Gross and Levitt seem to make this charge when they complain that feminist theorizing has "an unprecedented immunity to the scrutiny and skepticism that are standard for other fields of inquiry" (1994, p. 110). As I have shown earlier, it was Gross and Levitt who refused to scrutinize the feminist contributions to science; this is precisely the opposite result the feminist scientists in question hoped for and deserved.

Nevertheless, opponents of feminist approaches to the sciences rely heavily on an appeal to the invisible hand of the marketplace of scientific ideas: when they cannot get away with dismissing feminist scientific thought as "unscientific," or when the authors cannot be discredited as operating outside the sciences, feminist work must be characterized as an inevitable product of science-at-work. This appeal to the properly functioning scientific community does not sustain the conclusion drawn by opponents of feminist scientists, according to which the feminist origins of accepted scientific contributions are rendered irrelevant.

48 See, e.g., the passages by Levin, and by Gross and Levitt, quoted in Section 5.1.

49 Also Levin, p. 100.

Gross and Levitt acknowledge that the central issue at stake with feminist approaches to science is "the extent to which the prevalent feminist critique, as agent of methodological or conceptual change, is relevant to [science's] advance" (1994, p. 112). Now, consider the fact that Gross and Levitt help themselves to the following conclusions: "At times, baseless paradigms in medicine and the behavioral sciences have been pretexts for subordinating women. Pseudoscientific doctrines of innate inferiority and moral frailty have been used to discount female capacity for achievement and to confine women to subservient roles. All this is beyond dispute and generally recognized in intellectual circles" (1994, p. 110). We may expect a modest capitulation to the uses of feminist thought – at least in helping to unmask the baselessness of these paradigms and the "pseudo-" nature of these scientific doctrines – but no. Even though Gross and Levitt accept "that in scientific debate and in the process by which a preference for one paradigm over another emerges, attitudes of mind come into play that are in some measure dictated by social, political, ideological, and religious preconceptions," they maintain that it is the self-correcting dynamic of scientific method that does the real work: "Our reading of the history of science suggests . . . that theories leaning heavily on such props tend to be fragile and ephemeral, and that part of the increasing *power of scientific methodology* derives from always-increasing awareness of the danger that reasoning can be corrupted in this way if one is not careful [my emphasis]" (1994, p. 44). And finally, "We are *not* trying to deny that social interests and nonscientific belief systems often enter into the very human business of doing creative science," but "in the long run logic, empirical evidence, and explanatory parsimony are the masters . . . in the house of science" (1994, p. 56; their emphasis). This echoes Levin's claim that "the self-correcting character of the scientific method, with its emphasis on observation, the replication of experiments, and open discussion, insures that ['deviations from the ideal of objectivity'] will eventually be seen as such" (1988, p. 100).

The only gesture toward explaining how this process of self-correction actually functions lies in the claim that "If [scientific results] survive, they do so because they work, for a large number of people of hugely varied backgrounds and interests" (Gross and Levitt 1994, p. 112). Gross, Levitt, and Levin are unequipped to reckon with the details of how that self-correction – among people with "varied backgrounds and interests" – works: who offered the corrections and "increased the awareness," and why did they do so? How could such inadequate science have had such a long and influential run in the hands of the top scientists at the time?[50]

The heart of feminist analyses answers just these questions: including feminists among those with "varied backgrounds and interests" has and will contribute to the self-correction and advance of the sciences. Given that Gross and Levitt have already acknowledged the scientific importance of including different types of scientists, and that they admit that "expanding the pool of scientists will produce more and perhaps better science" (1994, p. 131),[51] they bear the burden of proof for concluding that feminist scientists are a type that ought either to be excluded, or to not count among the scientifically significant varied backgrounds and interests.

There is, thus, a striking logical problem for Gross, Levitt, and Levin's blanket dismissal of feminist contributions to the sciences, namely, they must launch and defend a rather demanding counterfactual: even if feminist scientists had not been the "correcting" force confronting this science, someone, sooner or later, would have provided such correction. In some cases, this is probably true; there is much common ground between feminist and other scientists' critiques of certain programs and explanations anyway (e.g., antireductionism, favoring models with higher complexity of interactions). But what is the significance of the claim that – even without the feminist contributions – the rest of the scientists would have eventually realized that something was wrong? I would remind readers that the critics' portrayal of the dynamics of self-correcting and objective sciences closely resembles the views articulated and advocated earlier by Longino, Fausto-Sterling, Harding, and other feminists. The only point of contention, it seems, lies in the counterfactual life of Gross, Levitt, and Levin, who appear to be making the petulant claim that even if feminists hadn't been there as participants in science's self-correction, everyone would have (eventually) gotten along fine without them. Whether or not this is so is completely irrelevant. Unless the critics can provide *reasons* – other than those that failed, above – that feminists *cannot* contribute to

[50] My analysis of the inconsistencies and grave evidential problems in recent evolutionary theorizing about women's orgasm has been met repeatedly with the response that it is "simply good science"; this reaction fails to engage the problem I address, namely, *why* it took decades for these able scientists to become *aware* that the evidence they cited undermined their own explanations. Lloyd (1993).

[51] They also admit that women's scientific "contributions have often in the past been undervalued" (1994, p. 123). How do they account for the fact that this happened in the first place? How do they account for the fact that it has, according to them, changed?

the usual processes of scientific inquiry, then their attempts to exclude feminists from the sciences amount to unvarnished dogmatism.

5.3. Exclusion at Any Price

Having evaluated the contradictions between some of the attacks on feminist approaches to science and their targets, we are faced with a mystery: if the feminists attacked by the critics for being anti-science or ignorant of science are actually neither, then what is going on here?

Many of the attacks involve the demonization that is essential to the identification of any group as essentially hostile, i.e., they embody the strategy of exclusion. Consider Gross and Levitt's description of feminists' "acute and apocalyptic oppositionism": "The announced goal, upon which feminists of the most disparate schools agree, is a science transformed . . . [this is an enormously] ambitious project: to refashion the epistemology of science from the roots up" (1994, pp. 32, 108). Similarly, we have Holton's improbable casting of a scientist and a philosopher of science as having an "alliance to a 'science' very different from conventional science," and as working toward "the end of science as we know it" (1993, pp. 152, 143).[52] This is all in the face of Longino's insistence that what she is urging as "feminist science" *"does not demand a radical break* with the science one has learned and practiced. The development of a 'new' science involves a more dialectical evolution with established science than the familiar language of scientific revolution implies" (Longino 1990, p. 193; my emphasis); as well as Sandra Harding's insistence, in the very article quoted by Holton, that she aims to restore "the ability of the sciences to provide objective empirically defensible descriptions and explanations of the regularities and underlying causal tendencies in nature and social relations" (1992, p. 1).[53] Note that the critics have misrepresented the feminist views in a particular fashion, one that attempts to distance the goals of feminists from recognizable and accepted goals of the sciences.

Under these circumstances, one might think it prudent for the critics to address themselves only to feminist analysts of science whose writings don't directly contradict their picture of feminist views. Instead, Gross

[52] The targets are Evelyn Fox Keller and Sandra Harding, respectively.

[53] She also states, "it's a very conservative notion of objectivity that I'm . . . proposing here . . . there are important aspects of the traditional notion of [scientific] objectivity which need not be challenged in order to accomplish the goals that I have in mind" (1992, p. 20).

and Levitt declare that "cultural constructivism is the underpinning of all these attacks, even when they are made by self-styled empiricists" (1994, p. 109; my emphasis). Now recall how Gross and Levitt attempted to show that Longino – in spite of her self-proclaimed attachment of empiricist goals – was *really* a card-carrying anti-scientist: they sifted out the substantive scientific criticism and pretended that it didn't exist; the remaining appeals to empirical standards were brushed off as ordinary or scientific common sense.[54]

Why would intelligent critics such as Gross and Levitt, who appear to have made efforts to read the relevant feminist works, place themselves in such a vulnerable position – one in which their frequent declarations of their "honest" evaluation of this literature (1994, pp. 113, 128, 241, 256)[55] are so blatantly violated – by their studied avoidance, combined with uncredited appropriation, of some of the strongest arguments offered by the feminists they claim to investigate? Gross and Levitt spare us from having to speculate about what they're really after: they want to *exclude* feminists from university and intellectual life, period.[56] Take a close look at this part of Gross and Levitt's description of "feminist criticism of science": they complain that "the best-known critics *are accepted as legitimate historians and philosophers of science, in circles far wider than their feminist peers*" (1994, p. 108; my emphasis). Here, we have the real story. Two of the researchers Gross and Levitt are most concerned to examine are Sandra Harding and Helen Longino, who were both trained as philosophers of science in highly ranked philosophy departments. Nevertheless, according to Gross and Levitt, we ought not consider historians and philosophers of science who have feminist views to be peers of other historians and philosophers of science; rather, the peer-group – the colleagues – of these scholars is to be defined by their political views. There could hardly be a clearer attempt to exclude participation by fully qualified professionals on the basis of ideology.

Gross and Levitt also repeatedly characterize feminist contributors to science as "nonscientists," and claim that what they have to say about science is naive and misguided, at best (1994, pp. 108, 122, 127, 159, 235, 236, 251). In addition to the problem, discussed above, that they bury, neglect,

[54] They make parallel moves versus Harding (1994, p. 249).

[55] Their editors should have advised Gross and Levitt that their avowal, "we are not dishonest," cannot help but recall Nixon's, "I am not a crook" (1994, p. 257).

[56] They argue, for example, that the high academic standing of these feminist researchers *itself* "raises serious questions about the presumed intellectual meritocracy of the academy" (Gross and Levitt 1994, p. 235).

or misrepresent the scientific nature of feminist contributions, Gross and Levitt put themselves in the awkward position of explicitly dismissing as scientifically incompetent their other two prime feminist targets: Evelyn Fox Keller, who has a Ph.D. in physics from Harvard and has produced laboratory results in molecular biology, and Donna Haraway, whose Ph.D. is in biology from Johns Hopkins.[57] How do the critics justify their apparent belief that holding feminist views actually, or effectively, subtracts or neutralizes even very high status scientific credentials?[58] I mention these cases to illustrate the stunningly arbitrary assignments of scientific competence made by Gross, Levitt, and Holton, depending on political or ideological affiliation. Such reassignments of scientific stature cannot be defended by any simplistic and unsophisticated appeal to an apparent conflict between holding political goals and the "objectivity" of science.

Finally, Gross and Levitt's book is peppered with phrases which suggest that feminist participation in scientific inquiry involves a category mistake. "Feminist science criticism" is represented as taking place in "women's studies programs," and they consistently refer to feminist writings on the science as "cultural" – rather than "scientific" – criticism. Both of these moves make it much easier for them to reach their conclusion that "the academic left's" *cultural* criticism doesn't make any real difference to practicing scientists.[59] To see the work being done by this misrepresentation, one only needs to consider their claim that: "most scientists are made aware of the academic left's critique only by fragmentary and sporadic contact"; how does this sound, when the truth is that feminists are themselves scientific colleagues in the same departments?

6. THE REAL ENEMIES OF OBJECTIVITY

I think that what motivates these books is fear – most significantly, fear of the loss of the sciences as authoritative resources for a host of social, political, and economic aims. In Section 2, I sketched some of the tensions and

[57] Although Holton clearly is referring to Keller's work, he perversely refuses to name her or to cite any of her books or articles (1993, p. 154).

[58] Gross and Levitt's inclusion of Haraway and Keller among the four chief representatives for feminist views of science belies their earlier aside that "a handful of figures with scientific credentials, as well as the occasional refugee from an unsatisfactory scientific career, can be found on the movement's fringes" (1994, p. 14; p. 6).

[59] "Science will not, in any serious way, be influenced, deflected, restricted, or even inconvenienced by these critics and those they influence" (Gross and Levitt 1994, p. 236; see pp. 3, 11, 112, 253–256).

conflicts that arise organically when the standards and ideals of scientific inquiry must be balanced with the interests of social and political control regarding the dissemination and publicity of information from and about the sciences. I have analyzed several of the most influential recent discussions and critiques that address these issues. Although these critiques may enhance awareness of the existence and shape of the tensions which I think are inherent in our democratic-cum-scientific society, I have found that their strategies for dealing with these conflicts are profoundly unsatisfactory on numerous counts.

On the more empirical side, it would seem that appropriate public consideration of their criticisms of social analyses of the sciences would involve careful examination of the relative evidential merits of each side. This would involve the critics in two difficult tasks: doing the work; and then engaging with the scholars they are attacking on their own turf. In some cases, the critics attempt to circumvent this necessity by transforming the appropriate questions and analytic frameworks of the fields of history, philosophy, and social sciences of science, in such a manner as to render them more tractable to the control of select natural scientists. Their other primary strategy involves equating feminist and social discussions of science with the rejection of objectivity, critical thought, and rationality. The fundamental problem with this strategy is that it involves *denying* that feminist contributors to science *share the pursuit of objective scientific knowledge and its rational applications in our world.*

The misrepresentations of feminist analyses, some of which I have documented in this paper, produce a double bind. Feminists have described some of the ways that aspects of the broader culture, which involve beliefs about sex and gender have historically shaped, and continue to have uncritical acceptance in, various aspects of scientific activities and products. Although they clearly mark their analyses and contributions as *adhering to and advancing the basic principles and standards guiding scientific thought,* they are nevertheless *treated as enemies* of the processes of scientific investigation. For example, one of the consistent themes of feminist discussions of the sciences is the importance of the full inclusion of qualified women – and others perceived as "feminine," such as racial minorities – in the development of scientific knowledge. The mass of arguments and evidence center around the value *to the sciences* of such inclusion. The cruel irony is that when these analyses of exclusion and arguments for the desirability – to the sciences – of inclusion are presented, they are not read as arguments for inclusion in social activities and processes that are *valued by feminists.* Instead, they are seen as *attacks by the enemies* of those pursuing scientific knowledge.

The significance of being labeled as "enemies" can become more visible through the analogy to military intelligence; such a label delineates a person's or group's relation to "inside" information. On the military model, the "sides" are clear, and one of the defining characters of the enemy is precisely "one who will use information against us" – this is why the controlling of information (including selective release of disinformation) is one of the cores of military strategy. Perhaps it will be easier to understand the level of alienation experienced by women and by feminists through recognizing that being denied full participation in producing analyses of scientific information – which they have earned the credentials to do – *amounts to being treated like "the enemy."* The urgency and vehemence which are sometimes seen in feminist critiques and analyses of the sciences ought to be understood in just this light; they *reflect* the depth of the hostility which is embodied by exclusion.

In other words, feminists object to being treated like "the enemy" when they are not. They object to having their scientific and scholarly credentials effectively taken away solely on account of their ideological interests, especially when they demonstrate, again and again, that they have *not* abandoned the fundamental standards and values of the scientific or philosophical communities in which they have earned full participation. Yet these very objections are also seen as further proof that they are the enemy. This is a vicious double bind.

There is an obvious conclusion to be drawn: those promoting the misrepresentations of feminist contributions do not want to see the feminists as friendly rather than hostile, perhaps because then they would have to include them; the goal *is* exclusion. The problem is that this goal is in direct conflict with the explicit ideologies of both scientific investigation and democratic society, each of which counts responsible participation as essential. The inherent conflicts that arise concerning the control and dissemination of information about the sciences can be resolved, pragmatically speaking, in a variety of ways, all of which include some degree of violation of the ideals of fairness, participation, or openness. The necessity of some level of pragmatic compromise, however, cannot justify the level and scope of misrepresentation and exclusion that are present in the critiques I have examined. And the reason is not simplemindedly moralistic, as the critics have contended. When social structures that rely for their functioning and success on ideologies of openness, fairness, and responsible participation rely instead – any more than they absolutely must – on secrecy, exclusion, and deception, *they cannot survive, much less thrive.*

The critics I have considered here are right to worry about defending the authority of the sciences against religious fundamentalism and superstition. Against science studies and feminism, however, they themselves stand as the true enemies of rationality, objectivity, and the long-term success of the sciences.

ACKNOWLEDGMENTS

I am especially grateful to Mathias Frisch, Jim Griesemer, David Hull, Matthew Nichter, Ina Roy, Michael Selgelid, and David Smith for their attention and criticism. Deepest thanks to Lynn Hankinson Nelson and Jack Nelson for their advice, encouragement, and patience.

12

Pre-Theoretical Assumptions in Evolutionary Explanations of Female Sexuality

My contribution to this Symposium focuses on the links between sexuality and reproduction from the evolutionary point of view.[1] The relation between women's sexuality and reproduction is particularly important because of a vital intersection between politics and biology – feminists have noticed, for more than a century, that women's identity is often defined in terms of her reproductive capacity. More recently, in the second wave of the feminist movement in the United States, debates about women's identity have explicitly included sexuality; much feminist argument in the late 1960s and early 1970s involved an attempt to separate out an autonomous female sexuality from women's reproductive functions.

It is especially relevant, then, to examine biological arguments, particularly evolutionary arguments, to see what they say about *whether* and *how* women's sexuality is related to reproduction. We shall find that many evolutionary arguments seem to support the direct linking of female sexuality and reproduction. Yet I will argue that this support is not well grounded. In fact, I think evolutionary explanations of female sexuality exemplify how social beliefs and social agendas can influence very *basic* biological explanations of fundamental physiological processes. In this paper, I shall spend some time spelling out a few examples in which assumptions about the close link between reproduction and sexuality yield misleading results, then I shall conclude with a discussion of the consequences of this case study for issues in the philosophy of science.

[1] This paper contains the text of an invited lecture delivered at the Symposium, Sex and Reproduction, Pacific Division APA, 1992. Further documentation and discussion will be found in *The Case of the Female Orgasm: Bias in the Science of Evolution* (Harvard University Press, 2005).

Lloyd, E. A. (1993). Pre-Theoretical Assumptions in Evolutionary Explanations of Female Sexuality. *Philosophical Studies*, *69*, 139–153.

The fundamental problem is that it is simply *assumed* that every aspect of female sexuality should be explained in terms of reproductive functions. But there is quite a bit of biological evidence that this is an empirically incorrect assumption to make. This raises the question of why autonomous female sexuality, distinct from reproductive functions, got left out of these explanations. I shall ultimately conclude that social context is playing a large and unacknowledged role in the practice of this science.

Perhaps the notion of the potential independence of female sexuality and reproduction may be unclear: I suggest thinking in terms of two distinct models, one in which all basic aspects of sexuality are *explained* in terms of reproduction, and the other in which sexuality is seen as an autonomous set of functions and activities, which are only *partially* explained in terms of reproductive functions. The difference may seem minor, but the two models have significantly disparate consequences when used in scientific explanation.

Let us begin with a classic and widespread model representing the hormonal determination of sexual behavior. In this model, female animals are only willing to have sexual intercourse when they are fertile – their sexual interest and activity are completely hormonally controlled. Typical and familiar examples of this type of setup include rats, dogs, and cats. When these animals are in estrus, they are willing and eager to mate, otherwise not. Technically, estrus is defined hormonally – that is, estrus is a particular phase of the menstrual cycle, in which the animal is fertile, and certain hormone measures are very high. This model embodies a *very tight* link between sexuality and reproduction: female sexuality functions completely in the service of reproduction.

Some interesting problems arise, however, in the application of this hormonally deterministic picture to human and nonhuman primate behavior.

First of all, although estrus is biologically defined as a hormonal state, it is very common for estrus to be defined *operationally* as the period in which "the female is willing to participate in sex." In one species, the bonobos, this behavioral definition led to the comic conclusion that this species is in estrus 57–86 percent of the time (Kano 1982). Notice that identifying estrus in this manner amounts to an *enforcement* of the belief that sexual behavior is tightly linked to reproduction. It becomes impossible even to *ask* whether these primates have an active sexual interest outside of their peak hormonal periods.

It turns out that when independent studies are made, sexual activity is not confined to the fertile phase for a number of nonhuman primates, including rhesus monkeys, several species of baboons, and common chimpanzees (Hafez 1971). Social factors such as partner preferences can be as influential as hormonal factors in regulating sexual behavior in several of these species (Wolfe 1979).

Female homosexual activity provides a good test for the assumed dependence of female sexuality on hormonal status. In addition, homosexual behaviors are clearly independent of reproduction *per se*, and might be interpreted as an indicator of an autonomous female sexuality. It turns out that female homosexual activities, which are widely observed in nonhuman primates, seem to be *independent* of the hormonal status of the participants. This independence has led some researchers to ignore such behaviors, or to declare that they are not, in fact, sexual. For example, pygmy chimpanzee females are commonly observed engaging in "genito-genital rubbing" (called "GG rubbing") in which two females hold each other and "swing their hips laterally while keeping the front tips of vulvae, where the clitorises protrude, in touch with each other" (Koruda 1980, p. 189). Kano argues that this behavior is not sexual, because nonhuman primates can only be "sexual" during estrus; the fact that pygmy chimps engage in GG rubbing outside of estrus, claims Kano, itself "suggests that this behavior does not occur exclusively in a sexual context, but has some other social significance" (Kano 1980, p. 243). Generally, some caution about the interpretation of apparently sexual behaviors is appropriate; the misunderstanding of many dominance behaviors as sexual ones plagued primatology in its first decades. At stake in this case, however, is the very *possibility* of hormonally independent female sexuality. The issue was resolved in 1984, when Mori, using a detailed study of statistical relations among behaviors, concluded that GG rubbing was, in fact, sexual behavior, because the same cluster of behavior surrounded both mating and GG rubbing (Mori 1984).

A more blatant example of researcher bias tying reproduction and female sexuality tightly together appears in an experiment being done on female orgasm in stumptail macaques. The original studies on female macaque orgasm, completed in the 1970s, documented female orgasm in the context of female homosexual mounting – that is, one female mounts another female, and stimulates herself to orgasm (Chevalier-Skolnikoff 1974). One very interesting result of these studies was the finding that the mounting, orgasmic female was *never* in estrus when these orgasms occurred. This is a provocative result for several reasons. First, according to the hormonal determinism model, female macaques are not supposed to be interested in any sexual activity outside of estrus. Second, these same female macaques *never* evidenced any sign of orgasm when they were participating in heterosexual coitus. A later study of the same species documented the same basic patterns, with the exception that four out of ten females in the group seemed, occasionally, to have orgasm during heterosexual coitus (Goldfoot, Westerborg-van Loon, Groeneveld, and Slob 1980).

I was surprised, therefore, when I spoke with a researcher who was working on the evolution of female orgasm in stumptail macaques.[2] He described his experimental setup to me with some enthusiasm: the females are radio-wired to record orgasmic muscle contractions and increased heart rate, and so on. This sounds like the ideal experiment, because it can record the sex lives of the females mechanically, without needing a human observer. In fact, the project had been funded by the NIH, and had presumably gone through the outside referee and panel reviews necessary for funding. But then the researcher described to me the clever way he had set up his equipment to record the female orgasms – he wired up the heart rate of the *male* macaques as the signal to start recording the *female* orgasms. When I pointed out that the vast majority of female stumptail orgasms occurred during sex among the females alone, he replied that yes, he knew that, but he was only interested in the *important* orgasms.

Obviously, this is a very unfortunate case. But it is not an isolated incident. Observations, measurements, interpretations, and experimental design are all affected by the background assumptions of the scientists. There is a pervasive and undefended assumption that female sexuality in non-human primates is tightly linked to reproduction. I would like now to explore briefly the situation regarding human beings.

HUMAN CASES

In most of the literature on the evolution of human sexuality, much attention is paid to the distinct attributes of human beings. The continual sexual "receptivity" of the human female is contrasted with the (supposed) strict hormonal restrictions on sexual activity in non-human animals. Human beings are supposed to be uniquely adapted to be sexually free from hormonal dictates, the possessors of a separate and self-constructed sexuality.

[2] The identity of this researcher is not included for publication. The information stated here was obtained through personal communication.

When it comes to evolutionary explanations of women's sexuality, though, the tight connection between reproduction and sexuality remains firmly in place.

To continue with the hormonal theme, we can begin by looking at beliefs about the distribution of female sexual interest during the menstrual cycle. Many researchers, in evolutionary biology, behavior, and physiology, have *deduced* that it must be the case in human females that peak sexual interest and desire occur at the same time as peak fertility. This conclusion is a simple extension of the hormonal determinism model from mice and dogs. Although this may have the ring of a reasonable assumption, it is not supported by the clinical literature. Kinsey, for example, found that 59 percent of his female sample experienced patterns of fluctuation in their sexual desire during their cycle – but only 11 percent experience a peak of sexual desire in mid-cycle, when they are most likely to be fertile (Kinsey et al. 1953). More recently, Singer and Singer, in a survey of studies, found that only 6–8 percent of women experience an increase in sexual desire around the time of ovulation. Most studies found peaks of sexual desire right before and after menstruation, when the woman is almost invariably infertile (Singer and Singer 1972).

Hence, the majority of evidence supports a picture in which female sexual interest and activity is clearly *decoupled* from her reproductive state. Sexual interest and motivation is highest when the woman is least likely to conceive. Unfortunately, a number of researchers working in the area of the evolution of sexuality have not taken this on board, and continue to assert that peak sexual desire *must* be around the time of ovulation – otherwise it would not make any sense.

This "making sense" is precisely what I'm interested in. According to these researchers, female sexuality doesn't *make sense* unless it is in the service of reproduction. There is no scientific defense offered for this assumption. A similar assumption is also present in the evolutionary explanations offered for female orgasm.

I have examined thirteen stories for the evolution of human female orgasm, and all except one of these stories assume that orgasm is an evolutionary adaptation. That is, they assume that orgasm conferred a *direct selective* advantage on its possessors, and that is how it came to be prevalent among women. The most common general formula for explaining the evolution of human female orgasm is through the pair-bond. Here, the pair-bond means more-or-less monogamous heterosexual coupling, and it is argued that such coupling increases the potential reproductive success of both parties through mutual cooperation and assistance with rearing offspring. The idea is that the male and the female in the pair bond provide mutual support to one another, and assist each other in rearing offspring, and that offspring raised under these conditions will tend themselves to have higher survival and reproductive success than those raised under other circumstances.

Hence, pair-bonding is seen as an adaptation in the evolutionary sense – it exists *because* it confers better chances of surviving and reproducing to those who display the trait. Under the assumption that pair-bonds are adaptive, frequent intercourse is also seen as adaptive, since it helps "cement the pair bond." And this is where orgasm comes in. Orgasm evolved, according to these pair-bond theorists, because it gave the female a reward and motivation to engage in frequent intercourse, which is itself adaptive, because it helps cement the pair bond. A number of different theorists have developed permutations of this basic story, but it remains the most widely accepted evolutionary story for female orgasm.[3]

Now, there is a glaring problem with this story – it assumes that intercourse is reliably connected to orgasm in females. All of the available clinical studies on women's sexual response indicate that this is a problematic assumption. Somewhere between 20 and 35 percent of women always or almost always experience orgasm with unassisted intercourse (Hite 1976; Kinsey et al. 1953 op cit.). I should add that this figure is supported by what cross-cultural information exists (see, e.g., Davenport 1977). This figure is very low, and it is especially striking given that somewhere around 90 percent of women do experience orgasm. Furthermore, about 30 percent of women *never* have orgasm with intercourse – this figure is taken from a population of women who do have regular intercourse, and of whom almost all are orgasmic (Hite 1976 op cit.). What this means is that *not* to have orgasm from intercourse is the experience of the majority of women the majority of the time. Not to put too fine a point on it, if orgasm is an adaptation that is a reward for engaging in frequent intercourse, it does not seem to work very well.

Obviously, this observation does not rule out the possibility that there is some selective advantage to female orgasm, but the salient point is that *none of these pair-bond theorists even address this problem*, which I

[3] Morris's work has been criticized by later researchers as being methodologically flawed (e.g., Wilson 1975; Crook 1972) but it is still widely cited, and its basic premises are accepted or slightly modified in other respected accounts such as Beach 1973; Pugh 1977; Crook 1972; Campbell 1967.

call the orgasm-intercourse discrepancy. Rather, they simply assume that when intercourse occurs, so does orgasm.[4]

In general, the association of intercourse with orgasm is relatively unproblematic among males. Hence, what is being assumed here is that female sexual response is like male sexual response to the same situation. There is little or no awareness, among the pair-bond theorists, of the orgasm-intercourse discrepancy, in spite of the fact that they cite or refer to the very studies which document this fact, including Kinsey's 1953 report on women's sexual response.

There is one obvious and understandable reason for this slip. They are, after all, trying to explain orgasm through evolutionary theory, which involves showing that the trait gave a reproductive advantage to its owner. It's easy to see how the equation of reproduction through intercourse and orgasm went by unnoticed. Nevertheless, this case does illustrate the main thesis, that female sexuality is unquestioningly equated with reproduction, and with the sort of sex that leads to reproduction.

There is another intriguing line of argument for the adaptive value of female orgasm, which was first published by Desmond Morris in 1967, although Shirley Strum tells me that Sherwood Washburn was teaching this in his classes at Berkeley earlier. Morris claimed that orgasm had a special function related to bipedalism (that is, walking on our hind legs), because it would increase chances of fertilization. Here again we have the direct link between female sexuality and reproduction.

> It does this in a rather special way that applies only to our own peculiar species. To understand this, we must look back at our primate relatives. When a female monkey has been inseminated by a male, she can wander away without any fear of losing the seminal fluid that now lies in the innermost part of her vaginal tract. She walks on all fours. The angle of her vaginal passage is still more or less horizontal. If a female of our own species were so unmoved by the experience of copulation that she too was likely to get up and wander off immediately afterwards, the situation would be different, for she walks bipedally and the angle of her vaginal passage during normal locomotion is almost vertical. Under the simple influence of gravity the seminal fluid would flow back down the vaginal tract and much of it would be lost. There is therefore . . . a great advantage in any reaction that tends to keep the female horizontal when the male ejaculates

[4] Typically, in evolutionary explanations, if a trait is taken to have evolved as an adaptation, yet is rarely used in the adaptive context, some explanation of the details of the selection pressure or the extreme adaptive value of the trait is offered.

> and stops copulation. The violent response of female orgasm, leaving the female sexually satiated and exhausted, has precisely this effect. (Morris 1967, p. 79)

Morris's view is in turn based on his understanding of physiological response – he says earlier . . . "after both partners have experienced orgasm [in intercourse] there normally follows a considerable period of exhaustion, relaxation, rest and frequently sleep" (Morris 1967, p. 55). Similarly, he claims, "once the climax has been reached, all the [physiological] changes noted are rapidly reversed and the resting, post-sexual individual quickly returns to the normal quiescent physiological state" (Morris 1967, p. 59).

Now let us refer to the clinical sex literature, which is cited by Morris and by others. According to this literature, the tendencies to states of sleepiness and exhaustion following orgasm, are, in fact, true for men but not for women. Regarding Morris's claim that the physiological changes are "rapidly reversed," this is also true for men but not for women – women return to the plateau phase of sexual excitement, and not to the original unexcited phase, as men do. This was one of the most noted conclusions of Masters and Johnson, whose picture of sexual response was enthusiastically adopted by Morris – but, it seems, only in part (Masters and Johnson 1966).

In fact, Masters and Johnson publicized an interesting and important difference between men's and women's sexuality, and that is the capacity of many women to have more than one orgasm without a significant break. Forty-seven percent of the women in Hite's survey did not feel that a single orgasm was always satisfying to them, and many women wanted more, some as many as fifteen to twenty five. If, at this point, you are concerned about Hite's bad reputation as a statistician and researcher, I'd like to point out that many of Hite's findings in that first study, published as the Hite report, were consistent with Kinsey's figures, and the Kinsey reports are considered, to this day, and in spite of any problems they might have, to be the best general studies ever done on the topic of women's sexuality (Kinsey 1953, pp. 375–376; Hite 1976, p. 417; Masters and Johnson 1966, p. 65). Masters and Johnson contrast the ability of many women to have five or six orgasms within a matter of minutes with the adult male's usual inability to have more than one orgasm in a short period (1961, p. 792). This female ability is linked to the fact that, following orgasm, women do not return to the prearoused state, as men do, but instead to the plateau phase of excitement.

Hence, Morris's story is in trouble. He claims that the physiological changes are rapidly reversed for women as well as for men. He also neglects the sizable percentage of women who are not satisfied by a single orgasm. Given the documented tendency in men to sleep and exhaustion following a single orgasm, it's not at all clear that a female desire to have orgasm wouldn't have exactly the opposite effect from that described by Morris – perhaps the woman would jump right up and cruise for a little more action at precisely the time when the sperm are most likely to leak out.

Actually, another serious problem with this story was recently pointed out by Shirley Strum, an expert on baboon behavior (personal communication). Supposedly, the selection pressure shaping female sexual response here is the potential loss of sperm that is threatened because human beings walk on two legs, and because the vaginal position is thus changed from horizontal to almost vertical. One would think, then, that our relatives walking on four legs would be protected against this occurrence, for anatomical reasons. But Strum says that immediately following intercourse, female baboons like to go off and *sit down* for ten or fifteen minutes. When they get up, she says, they inevitably leave a visible puddle of semen on the ground. Perhaps, then, the loss of semen is not the serious evolutionary challenge that Desmond Morris and others take it to be.

SUMMARY

I claim that social agendas appear in these stories through the obliteration of any female sexual response that is independent from her function as a reproducer. Autonomous, distinct female sexual response just disappears.

In these explanations women are presumed to have orgasms nearly always with intercourse, as men do. Women are presumed to return to the resting state following orgasm, as men do. One could object that Morris is a relatively easy target, so I will offer the following tidbit in defense of my analysis. Gordon Gallup and Susan Suarez published, in 1983, a technical discussion on optimal reproductive strategies for bipedalism, and took up Morris's antigravity line of argument. They argue that orgasm would be adaptive because it would keep the woman lying down, and hence keep the semen from escaping. In the context of these paragraphs on female orgasm, they state, "it is widely acknowledged that intercourse frequently acts as a mild sedative. The average individual requires about five minutes of repose before returning to a normal state after orgasm"

(Gallup and Suarez 1983, p. 195). The scientific reference they offer for this particular generalization is Kinsey (1948), which is, in fact, exclusively on *male* sexual response. In other words, this "average individual," which figures in their story about female orgasm, is, in fact, explicitly male.

AN ALTERNATIVE EXPLANATION

Donald Symons, in his book *The Evolution of Human Sexuality* (1979), argues that female orgasm is not an adaptation. He develops a story parallel to the one about male nipples – female orgasm exists because orgasm is strongly selected in males, and because of their common embryological form, women are born with the potential for having orgasms, too.[5] Part of the story, then, is that orgasm is strongly selected in males; this is fairly plausible, since it is difficult for male mammals to reproduce without ejaculation, which requires a reflex response in certain muscles. These muscles are, in fact, the same (homologous) muscles that are involved in female orgasm. It is also significant that the intervals between contractions in orgasm is four-fifths of a second in both men and women. This is considered evidence that orgasm is a reflex with the same developmental origin in both sexes.

One of the consequences of Symons's theory is that it would be expected that similar stimulation of the clitoris and penis would be required to achieve the same reaction or reflex response. This similarity shows especially in the figures on masturbation. Only 1.5 percent of women masturbate by vaginal entry, which provides stimulation similar to the act of intercourse; the rest do so by direct or indirect stimulation of the clitoris itself (Kinsey 1953; Hite 1976, pp. 410–411). Also, on the developmental theory, one would *not* expect similar reactions to intercourse, given the differences in stimulation of the homologous organs.

Finally, this theory is also supported by the evidence of orgasm in nonhuman primates. The observed orgasms occur almost exclusively when the female monkeys are themselves mounting other monkeys, and not during copulation. On the nonadaptive view of orgasm, this is almost to be expected. There, female orgasm is defined as a potential, which, if the female gets the right sort and amount of stimulation, is activated. Hence, it is not at all surprising that this does not occur often during copulation,

[5] This argument is spelled out in more detail by Stephen Jay Gould (1987), in an essay that was based on my research and arguments.

which in these monkeys includes very little, if any, stimulation of the clitoris, but occurs rather with analogous stimulation of the homologous organs that they get in mounting.

Symons's proposal, which I found very powerful and plausible, has been sharply criticized by a number of feminists. For instance, a leading feminist sociobiologist, Sarah Blaffer Hrdy, claims that this non-adaptive explanation is dismissive of female sexuality (1981, p. 165). Similarly, Mina Caulfield (1985) accuses Symons of denying the "significance of female pleasure."

I view these criticisms as misguided, because they are based on the assumption that *only* adaptive explanations can provide for the significance of a trait. But why should we believe this? Musical and singing ability are not adaptations, but they are very important to human culture and human life. One must have adopted the idea, not merely that "what is natural is good," but, furthermore, that "only what is adaptive is good." The evolutionarily derivative role of female orgasm implies absolutely nothing about its importance unless you are a committed adaptationist. Finally, I wonder why these feminists are so eager to get orgasm defined as an adaptation – several of the serious evidential problems with evolutionary explanations about female orgasm arose, I have argued, from making an easy connection between sexuality and reproduction.

I would like to just mention a possible alternative interpretation. The conclusion that orgasm is not an adaptation *could* be interpreted as emancipatory. After all, the message here is that orgasm is a freebie. It can be used in any way that people want; there is no "natural" restriction on female sexual activities, nor is there any scientific ground for such a notion. Under the developmental view, the constraints are loosened on possible explanations about women's sexuality that are consistent with accepted clinical conclusions and with evolutionary theory. Hence, the realm formerly belonging exclusively to reproductive drive would now be open to much, much more.

DISCUSSION

I would like to draw two conclusions.

First, I believe that prior assumptions have more influence in these areas of science than is commonly acknowledged in the usual philosophical and scientific pictures of scientific theorizing and testing. In the cases examined here, science is not very separate from the social and cultural context. Rather, social assumptions and prior commitments of the scientists play a major role in the practice of science itself, at many levels – experimental design, data collection, predictions, hypothesis formulation, and the evaluation of explanations.

To understand this area of scientific practice, we need a view of science that is more sophisticated, one that has more moving parts, than the pictures typically presented by philosophers of science. Under the usual approaches, science is seen as involving relations purely between theory and data, or between theory, data, and explainer. But this is not enough. We need a way to recognize and analyze the vital role of pre-theoretical beliefs and categories in *all* stages of scientific research.

One might object that the subject matter of this part of science makes social influence inevitable, and that one would not expect this same level of cultural bias in other scientific contexts. That's probably right. But we do not need to show social forces at work in every possible case of scientific inquiry in order to insist on having a theory of science with enough flexibility to work in many areas. The cases I have presented here are definitely "science," with plenty of funding, backing, authority, influence, and prestige. Philosophers who insist on a *pure view* of science, based on isolated and idealized examples of physics, are voting themselves out of the action. There are very interesting and important things going on in other areas as well, as the cases I have outlined earlier attest. Developing a view of science that can account for these other fields is vital.

My suggestion does *not* involve commitment to a relativist position. In a complete analysis of evolutionary explanations of human sexuality, I would adopt Helen Longino's general approach, in which she characterizes objectivity in science as resulting from the critical interaction of different groups and individuals with different social and cultural assumptions and different stakes. Under this view, the irreducibility of the social components of the scientific situation is accounted for – these social assumptions are, in fact, an essential part of the picture of scientific practice.

At any rate, I take it that the cases I have described above violate our common philosophical understandings of how we arrive at scientific beliefs, how knowledge is created, and how science works. If philosophers go the route of labeling as "science" *only* that which obeys the demands of current philosophy, we will end up discussing only some parts of physics and maybe some math. Meanwhile, what about the rest of science – biology, social sciences, anthropology, psychology, biochemistry? I suggest adopting and developing recent contextualist and feminist views of science, which take explicit account of pre-theoretical assumptions and preconceptions, and their social origins.

This case involving female sexuality is very interesting because there are *two* very strong forces working to put sex and reproduction together. Adaptationism, within biology, promotes the easy linking of all sexual activity with reproduction success, the measure of relative fitness. Second, the long social tradition of *defining* women in terms of their sexual and reproductive functions alone also tends to link sexuality and reproduction more tightly than the evidence indicates.

The long struggle by various women's movements to separate sex and reproduction seems to have had very little effect on the practice of the science we have examined in this paper. This is especially ironic, because politically, ever since the late Nineteenth Century, scientific views about gender differences and the biology of women have been the single most powerful political tool against the women's movements. My second and more controversial conclusion is that current "purist" philosophy of science actually *contributes to* that political power by reinforcing myths of the insulation of scientific endeavors from social influences. A more sophisticated understanding of the production and evaluation of scientific knowledge would mean seeing science as (partly) a continuation of politics. Science would then lose at least *some* independent authority in the political arena. Judging by the scientific work that I have discussed in this paper, I think that would be a good thing.

References

Alcoff, L., and E. Potter (Eds.). (1993). *Feminist Epistemologies*. New York: Routledge.

Altmann, J. (1974). Observational Study of Behavior: Sampling Methods. *Behavior, 49*, 227–267.

Angier, N. (1991, April 14). Molecular "Hot Spot" Hits at a Cause of Liver Cancer. *New York Times*.

Antony, L. (1993). Quine as Feminist: The Radical Import of Naturalized Epistemology. In L. Antony and C. Witt (Eds.), *A Mind of One's Own: Feminist Essays on Reason and Objectivity*. Boulder, CO: Westview Press.

Antony, L., and C. Witt (Eds.). (1993). *A Mind of One's Own: Feminist Essays on Reason and Objectivity*. Boulder, CO: Westview Press.

Aoki, K. (1982). A Condition for Group Selection to Prevail over Counteracting Individual Selection. *Evolution, 36*, 832–842.

Arnold, A. J., and K. Fristrup (1982). The Theory of Evolution by Natural Selection: A Hierarchical Expansion. *Paleobiology, 8*, 113–129.

Arnold, S. J., and M. J. Wade (1984). On the Measurement of Natural and Sexual Selection: Applications. *Evolution, 38*, 720–734.

Atkinson, T. (1974). *Amazon Odyssey*. New York: Links Books.

Awards: 1989. (1989). *Science, 243*, 672 (3).

Axelrod, R. (1984). *The Evolution of Cooperation*. New York: Basic Books.

Axelrod, R., and W. D. Hamilton (1981). The Evolution of Cooperation. *Science, 211*, 1390–1396.

Babbitt, S. E. (1993). Feminism and Objective Interests: The Role of Transformation Experiences in Rational Deliberation. In L. Alcoff and E. Potter (Eds.), *Feminist Epistemologies*. New York: Routledge.

Bailey, N. T. J. (1967). *The Mathematical Approach to Biology and Medicine*. London and New York: Wiley.

Baird, P. A. (1990). Genetics and Health Care: A Paradigm Shift. *Perspectives in Biology and Medicine, 33*, 203–213.

Barkow, J., L. Cosmides, and J. Tooby (1992). *The Adapted Mind*. New York: Oxford University Press.

Bayer, R. (1981). *Homosexuality and American Psychology: The Politics of Diagnosis*. New York: Basic Books.
Beach, F. (1973). Human Sexuality and Evolution. In W. Montanga and W. Sadler (Eds.) *Advances in Behavioral Biology* (pp. 333–365). New York: Plenum Press.
Beatty, J. (1980). Optimal-Design Models and the Strategy of Model Building in Evolutionary Biology. *Philosophy of Science, 47*, 532–561.
Beatty, J. (1981). What's Wrong with the Received View of Evolutionary Theory? *PSA 1980: Vol. Two*. East Lansing, MI: Philosophy of Science Association.
Beatty, J. (1982). The Insights and Oversights of Molecular Genetics: the Place of the Evolutionary Perspective. *PSA 1982: Vol. One*. East Lansing, MI: Philosophy of Science Association.
Beckner, M. (1959). *The Biological Way of Thought*. New York: Columbia University Press.
Bennett, D. (1975). The T-Locus of the Mouse. *Cell, 6*, 441–454.
Bennett, J. (1965). Substance, Reality, and Primary Qualities. *American Philosophical Quarterly, 2*(1), 1–17.
Berkeley, G. (1871). *Works*. Oxford: Oxford University Press.
Bernstein, R. (1983). *Beyond Objectivism and Relativism*. Philadelphia: University of Pennsylvania Press.
Bernstein. R. (1988). The Rage against Reason. In E. McMullin (Ed.), *Construction and Constraint: The Shaping of Scientific Rationality*. South Bend, IN: University of Notre Dame Press.
Bleier, R. (1984). *Science and Gender: A Critique of Biology and Its Theories on Women*. New York: Pergamon Press.
Bleier, R. (1986). Sex Differences Research: Science or Belief? In R. Bleier (Ed.), *Feminist Approaches to Science*. New York: Pergamon.
Blonder, L. (1993). Review of *The Adapted Mind. American Anthropologist, 95*, 777–778.
Bock, W. (1980). The Definition and Recognition of Biological Adaptation. *American Zoologist, 20*, 217–27.
Boorman, S. A., and P. R. Levitt (1973). Group Selection on the Boundary of a Stable Population. *Theoretical Population Biology, 4*, 85–128.
Boorman, S. A. (1978). Mathematical Theory of Group Selection: Structure of Group Selection in Founder Populations Determined from Convexity of the Extinction Operator. *Proceedings of the National Academy of Sciences, USA, 69*, 1909–13.
Bordo, S. (1987). *The Flight to Objectivity: Essays on Cartesianism and Culture*. Albany: State University of New York Press.
Bordo, S. (1989). The View from Nowhere and the Dream of Everywhere: Heterogeneity, Adequation and Feminist Theory. *American Philosophical Association Newsletter on Feminism and Philosophy, 88*(2), 19–25.
Boyd, R. (1980). The Current Status of Scientific Realism. *Proceedings of the Philosophy of Science Association*. East Lansing, MI: Philosophy of Science Association.
Boyle, R. (1744). *The Works of the Honourable Robert Boyle*. London.
Brandon, R. N. (1978). Adaptation and Evolutionary Theory. *Studies in History and Philosophy of Science, 9* (3), 181–206.

Brandon, R. N. (1981a). Biological Teleology: Questions and Explanations. *Studies in History and Philosophy of Science, 12* (2), 91–105.

Brandon, R. N. (1981b). A Structural Description of Evolutionary Theory. *Philosophy of Science Association, 2*, 427–439.

Brandon, R. N. (1982). The Levels of Selection. *Proceedings of the Philosophy of Science Association 1982*, 1, 315–23.

Brandon, R. N. (1985). Adaptation Explanations: Are Adaptations for the Good of Replicators or Interactors? In D. Depew and B. Weber (Eds.), *Evolution at a Crossroads: The New Biology and the New Philosophy of Science* (pp. 81–96). Cambridge, MA: MIT Press/Bradford.

Brandon. R. N. (1988). Levels of Selection: A Hierarchy of Interactors. In H. C. Plotkin (Ed.), *The Role of Behavior in Evolution* (pp. 51–71). Cambridge, MA: MIT Press.

Brandon, R. N. (1990). *Adaptation and Environment.* Princeton, NJ: Princeton University Press.

Bruck, D. (1957). Male Segregation Ratio Advantage as a Factor in Maintaining Lethal Alleles in Wild Populations of House Mice. *Proceedings of the National Academy of Sciences USA, 43*, 152–158.

Bum, E. A. (1932). *The Metaphysical Foundations of Modern Physical Science.* Garden City, NY: Doubleday/Anchor.

Burian, R. M. (1983). Adaptation. In M. Greene (Ed.), *Dimensions of Darwinism* (pp. 287–314). Cambridge: Cambridge University Press.

Butts, R. E. (1977). Consilience of Inductions and the Problem of Conceptual Change in Science. In *Logic, Laws and Life*. Pittsburgh: University of Pittsburgh Press.

Callebaut, W. (Ed.) (1993). *The Naturalistic Turn: How Real Philosophy of Science Is Done.* Chicago: University of Chicago Press.

Campbell, B. (1967). *Human Evolution: An Introduction to Man's Adaptations.* Chicago: Aldine.

Caplan, A. L. (1981). The Concept of Health and Disease. In R. M. Veatch, (Ed.) *Medical Ethics* (pp. 49–62). Boston: Jones & Bartlett.

Carnap, R. (1928). *Der Logische Aufbau der Welt.* Berlin-Schlachtensee: Weltkreis-Verlag.

Carnap, R. (1950). Empiricism, Semantics, and Ontology. *Revue Internationale de Philosophie, 4*, 20–40.

Carnap, R. (1956). *Meaning and Necessity: A Study in Semantics and Modal Logic.* Chicago: University of Chicago Press.

Carnap, R. (1967). *The Logical Structure of the World: Pseudoproblems in Philosophy*. Berkeley: University of California Press.

Cartwright, N. (1983). *How the Laws of Physics Lie.* Oxford: Oxford University Press.

Cartwright, N. (1990). *Nature's Capacities and Their Measurement.* Oxford: Oxford University Press.

Cassidy, J. (1978). Philosophical Aspects of the Group Selection Controversy. *Philosophy of Science, 45*, 575–594.

Cat, J. (1998). The Physicists' Debates on Unification in Physics at the End of the 29th Century. *Historical Studies of the Physical Sciences, 28* (2), 253–299.

Cat, J. (2000). Must Microcausality Condition Be Interpreted Causally? Beyond Reduction and Matters of Fact. *Theoria, 37*, 59–85.

Cat, J. (forthcoming). *Physics Beyond Laws & Theories: The Limits of Unity, Universality and Precision*. Baltimore, MD: Johns Hopkins University Press.

Cheetham, A. H. (1986). Tempo of Evolution in a Neogene Bryozoan: Rates of Morphologic Change Within and Across Species Boundaries. *Paleobiology, 12*, 190–202.

Cheng, P. W., and K. J. Holyoak (1983). Schema-Based Inferences in Deductive Reasoning. Anaheim, 25th Annual Meeting of the American Psychological Association.

Cheng, P. W., and K. J. Holyoak (1984). Pragmatic Schemas for Deductive Reasoning. San Antonio, 25th Annual Meeting of the Psychonomic Society.

Cheng, P. W., and K. J. Holyoak (1985). Pragmatic Reasoning Schemas. *Cognitive Psychology, 17*, 391–416.

Cheng, P. W., K. Holyoak et al. (1986). Pragmatic Versus Syntactic Approaches to Training Deductive Reasoning. *Cognitive Psychology, 18*, 293–328.

Cheng, P. W., and K. J. Holyoak (1989). On the Natural Selection of Reasoning Theories. *Cognition, 33*, 285–313.

Chevalier-Skolnikoff, S. (1974). Male-Female, Female-Female, and Male-Male Sexual Behavior in the Stumptail Monkey, with Special Attention to the Female Orgasm. *Archives of Sexual Behavior, 3*(2), 95–116.

Chevalier-Skolnikoff, S. (1976). Homosexual Behavior in a Laboratory Group of Stumptail Monkeys (Macaca Arctoides): Forms, Contexts, and Possible Social Functions. *Archives of Sexual Behavior, 5*(6), 511–527.

Chomsky, N. (1975). *Reflections on Language*. New York: Random House.

Chomsky, N. (1980). *Rules and Representations*. New York: Columbia University Press.

Churchland, P. M. (1989). *A Neurocomputational Perspective*. Cambridge, MA: MIT Press.

Churchland, P. S. (1986). *Neurophilosophy*. Cambridge, MA: MIT Press.

Churchland, P. S. (1993). Can Neurobiology Teach Us Anything about Consciousness? *Proceedings of the American Philosophical Association, 67*(4), 23–39.

Code, L. (1991). *What Can She Know? Feminist Theory and the Construction of Knowledge*. Ithaca, NY: Cornell University Press.

Code, L. (1993). Taking Subjectivity into Account. In L. Alcoff and E. Potter (Eds.), *Feminist Epistemologies* (pp. 15–48). New York: Routledge.

Colwell, R. K. (1981). Evolution of Female-Based Sex Ratios: The Essential Role of Group Selection. *Nature, 290*:401–4.

Conkey, M. W. (1984). Archaeology and the Study of Gender. *Advances in Archaeological Method and Theory, 7*, 1–38.

Conkey, M. W., and S. H. Williams (1991). Original Narratives: The Political Economy of Gender in Archaeology. In M. di Leonardo (Ed.), *Gender at the Crossroads of Knowledge: Feminist Anthropology in the Postmodern Era*. Berkeley: University of California Press.

Cooper, W. S. (1984). Expected Time to Extinction and the Concept of Fundamental Fitness. *Journal of Theoretical Biology, 107*, 603–629.
Cosmides, L., and J. Tooby (1987). From Evolution to Behavior: Evolutionary Psychology as the Missing Link. In J. Dupre (Ed.), The *Latest on the Best: Essays on Evolution and Optimality* (pp. 277–306). Cambridge, MA: MIT Press.
Cosmides, L., and J. Tooby (1995). Beyond Intuition and Instinct Blindness: Toward an Evolu-tionary Rigorous Cognitive Science. In J. Mehler and S. Franck (Eds.), *Cognition on Cognition* (pp. 69–105). Cambridge, MA: MIT Press.
Cosmides, L. (1985). Deduction or Darwinian Algorithms?: An Explanation of the "Elusive" Content Effect on the Wason Selection Task. Ph.D. Dissertation, Harvard University.
Cosmides, L. (1989). The Logic of Social Exchange: Has Natural Selection Shaped How Humans Reason? Studies with the Wason Selection Task. *Cognition, 31*, 187–276.
Council for Responsible Genetics. (1990). *Position Paper on Human Genome Initiative*. Boston: Committee for Responsible Genetics.
Craig, D. M. (1982). Group Selection versus Individual Selection: An Experimental Analysis. *Evolution, 36*, 271–82.
Cranor, C. F. (1994). Genetic Causation. In C. F. Cranor, (Ed.), *Are Genes Us?* (pp. 125–141). New Brunswick, NJ: Rutgers University Press.
Crook, J. H. (1972). Sexual Selection, Dimorphism, and Social Organization in the Primates. In B. Campbell (Ed.), *Sexual Selection and the Descent of Man*. Chicago: Aldine.
Cronin, H. (1991). *The Peacock's Tail*. Oxford: Oxford University Press.
Crow, J. F., and K. Aoki (1982). Group Selection for a Polygenetic Behavioral Trait: A Differential Proliferation Model. *Proceedings of the National Academy of Sciences, USA, 79*, 2628–2631.
Crow, J. F., and M. Kimura (1970). *An Introduction to Population Genetics*. New York: Harper and Row.
Culliton, B. J. (1990). Mapping Terra Incognita (Humani Corporis). *Science, 250*, 211.
Curley, E. (1972). Locke, Boyle, and the Distinction between Primary and Secondary Qualities. *Philosophical Review, 81*, 438–464.
Curley, E. (1978). *Descartes against the Skeptics*. Cambridge, MA: Harvard University Press.
Damuth, J., and Heisler, I. L. (1988). Alternative Formulations of Multilevel Selection. *Biology and Philosophy, 3*, 407–30.
Darden, L., and N. Maull (1977). Interfield Theories. *Philosophy of Science, 44*(1), 43–64.
Darlington, C. D. (1939). *The Evolution of Genetic Systems*. Cambridge: Cambridge University Press.
Darwin, C. (1903). *More Letters of Charles Darwin* (Francis Darwin, Ed.). New York: D. Appleton.
Darwin, C. (1919). *Life and Letters of Charles Darwin* (Francis Darwin, Ed.). New York: D. Appleton.

Darwin, C. (1964). *On the Origin of Species (1st edition facsimile)*. Cambridge, MA: Harvard University Press.

Darwin, C. (1967). *Darwin and Henslow: The Growth of an Idea* (Nora Barlow, Ed.). Berkeley and Los Angeles: University of California Press.

Daston, L. (1992). Objectivity and the Escape from Perspective. *Social Studies of Science, 22*, 597–618.

Daston, L., and P. Galison (1992). The Image of Objectivity. *Representations, 40*, 81–128.

Davenport, W. (1977). Sex in Cross-Cultural Perspective. In F. Beach (Ed.), *Human Sexuality in Four Perspectives* (pp. 115–163). Baltimore, MD: Johns Hopkins University Press.

Davidson, D. (1967). Truth and Meaning. *Synthese, 17*, 304–323.

Davidson, D. (1974). The Very Idea of a Conceptual Scheme. *Proceedings and Addresses of the American Philosophical Association, 47*, 5–20.

Davis, B. D. (1990). The Human Genome and Other Initiatives. *Science, 249*, 342.

Dawkins, R. (1978). Replicator Selection and the Extended Phenotype. *Zeitschrill fur Tierpsychologie, 47*, 61–76.

Dawkins, R. (1982a). Replicators and Vehicles. *In King's College Sociobiology Group, Cambridge, Current Problems in Sociobiology* (pp. 45–64). Cambridge: Cambridge University Press.

Dawkins, R. (1982b). *The Extended Phenotype*. New York: Oxford University Press.

Dawkins, R. (1986). *The Blind Watchmaker*. New York: Norton.

Dawkins, R. (1989a). The evolution of evolvability. In C. Langdon (Ed.), *Artificial Life, Santa Fe Institute Studies in the Sciences of Complexity* (pp. 201–220). Reading, MA: Addison-Wesley.

Dawkins, R. (1989b). *The Selfish Gene*, Revised Edition. New York: Oxford.

de Waal, F. B. M. (1989). *Peacemaking among Primates*. Cambridge, MA: Harvard University Press.

de Waal, F. B. M. (1991a). The Chimpanzee's Sense of Social Regularity and Its Relation to the Human Sense of Justice. *American Behavioral Scientist, 34*(3), 335–349.

de Waal, F. B. M. (1991b). Complementary Methods and Convergent Evidence in the Study of Primate Social Cognition. *Behaviour, 118*, 297–320.

de Waal, F. B. M., and A. H. Harcourt (Eds). (1992). *Coalitions and Alliances in Humans and Other Animals*. New York: Oxford University Press.

de Waal, F B. M., and D. L. Johanowicz (1993). Modification of Reconciliation Behavior through Social Experience: An Experiment with Two Macaque Species. *Child Development, 64*, 897–908.

de Waal, F. B. M., and L. M. Luttrell (1988). Mechanisms of Social Reciprocity in Three Primate Species: Symmetrical Relationship Characteristics or Cognition? *Ethology and Sociobiology, 9*, 101–118.

Dennett, D. (1978a). *Brainstorms*. Cambridge, MA: MIT Press.

Dennett, D. (1978b). Current Issues in the Philosophy of Mind. *American Philosophical Quarterly, 15*(4), 249–261.

Dennett, D. (1988). Quining Qualia. In A. J. Marcel and E. Bisiach (Eds.), *Consciousness in Contemporary Science*. Clarendon, UK: Oxford University Press.
Dennett, D. (1991). *Consciousness Explained*. Boston, MA: Little, Brown and Co.
Dewey, J. (1925). *Experience and Nature*. La Salle, IL: Open Court Publishing.
Dewey, J. (1929). *The Quest for Certainty*. New York: Minton, Balch.
Dewey, J. (1938). *Logic: The Theory of Inquiry*. New York: Henry Holt.
Diamond, J. M. (1969). Avifaunal Equilibria and Species Turnover Rates on the Channel Islands of California. *Proceedings of the National Academy of Sciences, USA, 64*, 57–63.
Dobzhansky, T. (1937) *Genetics and the Origin of Species*. New York: Columbia University Press.
Dobzhansky, T. (1956). What Is an Adaptive Trait? *American Naturalist, 40* (855), 337–347.
Dobzhansky, T. (1968). Adaptedness and Fitness. In R. C. Lewontin (Ed.), *Population Biology and Evolution*, (p. 109–21). Syracuse, NY: Syracuse University Press.
Dobzhansky, T. (1970). *Genetics of the Evolutionary Process*. New York: Columbia University Press.
Dugatkin, L. A., and H. K. Reeve (1994). Behavioral Ecology and Levels of Selection: Dissolving the Group Selection Controversy. *Advances in the Study of Behavior, 23*, 101–133.
Dulbecco, R. (1986). A Turning Point in Cancer Research: Sequencing the Human Genome. *Science, 231*, 1055–1056.
Dunbar, R. I. M. (1982). Adaptation, Fitness and the Evolutionary Tautology. In *King's College Sociobiology Group, Cambridge, Current Problems in Sociobiology* (pp. 9–28). Cambridge: Cambridge University Press.
Dupre, J. (1983). The Disunity of Science. *Mind, 92*, 321–346.
Dupre, J. (1993). *The Disorder of Things*. Cambridge, MA: Harvard University Press.
Duran, J. (1991). *Toward a Feminist Epistemology*. Savage, MD: Rowman & Littlefield.
Economist. (1993). Review of *The Adapted Mind*. *326*, 82.
Eldredge, N. (1971). The Allopatric Model and Phylogeny in Paleozoic Invertebrates. *Evolution, 25*, 156–167.
Eldredge, N. (1985). *Unfinished Synthesis: Biological Hierarchies and Modern Evolutionary Thought*. New York: Oxford University Press.
Eldredge, N., and S. J. Gould (1972). Punctuated Equilibrium: An Alternative to Phyletic Gradualism. In T. J. M. Schopfied, (Ed.), *Models in Paleobiology*. San Francisco: Freeman.
Fausto-Sterling, A. (1985). *Myths of Gender: Biological Theories about Women and Men*. New York: Basic Books.
Fausto-Sterling, A. (1993). The Five Sexes: Why Male and Female Are Not Enough. *The Sciences, 33*(2), 20–24.
Fedigan, L. (1982). *Primate Paradigms*. Montreal, Can: Eden.
Feyerabend, P. (1975). *Against Method*. London: New Left Bookstore.

Fine, A. (1984). *The Natural Ontological Attitude.* In *Scientific Realism.* Berkeley: University of California Press.
Fine, A. (1986). *The Shaky Game: Einstein, Realism and the Quantum Theory.* Chicago: University of Chicago Press.
Fisher, R. A. (1930). *The Genetical Theory of Natural Selection.* London: Oxford University Press.
Fisher, R. A. (1958). *The Genetical Theory of Natural Selection.* New York: Dover.
Fiske, A. P. (1991). *Structures of Social Life: The Four Elementary Forms of Human Relations.* New York: Free Press.
Fodor, J. A. (1983). *The Modularity of Mind: An Essay on Faculty Psychology.* Cambridge, MA: MIT Press.
Foucault, M. (1972). *The Archaeology of Knowledge.* New York: Pantheon.
Foucault, M. (1987). Interview of M. Foucault: Questions of Method. In *After Philosophy: End or Transformation?* Cambridge, MA: MIT Press.
Fowler, C. W., and J. A. MacMahon (1982). Selective Extinction and Speciation: Their Influence on the Structure and Functioning of Communities and Ecosystems. *American Naturalist, 119,* 480–498.
Fradkin, P. (1989). *Fallout.* Tucson: University of Arizona Press.
Franklin, I., and R. C. Lewontin (1970). Is the Gene the Unit of Selection? *Genetics, 65,* 707–734.
Friedman, T. (1989). Progress toward Human Gene Therapy. *Science, 244,* 1275.
Friedman, T. (1990). The Human Genome Project – Some Implications of Extensive "Reverse Genetic" Medicine [opinion]. *American Journal of Human Genetics, 46,* 409.
Frieze, I. H., J. E. Parsons, P. B. Johnson, D. N. Ruble, and G. L. Zellman (1978). *Women and Sex Roles: A Social Psychological Perspective.* New York: W. W. Norton.
Fujimara, J. (1998). The Molecular Biological Bandwagon in Cancer Research: Where Social Worlds Meet. *Social Problems, 35,* 261.
Gajdusek, D. C. (1970). Physiological and Psychological Characteristics of Stone Age Man. *Engineering and Science, 22,* 56–62.
Gallup, G. G., and S. D. Suarez (1983). Optimal Reproductive Strategies for Bipedalism. *Journal of Human Evolution, 12,* 195.
Garber, D. (1978). Science and Certainty in Descartes. In *Descartes: Critical and Interpretive Essays.* Baltimore, MD: Johns Hopkins University Press.
Garber, D. (1992). *Descartes' Metaphysical Physics.* Chicago: University of Chicago Press.
Garber, D. (1993). Descartes and Experiment in the Discourse and Essays. In *Essays on the Philosophy and Science of Rene Descartes.* New York: Oxford University Press.
Geertz, C. (1990, November 8). A Lab of One's Own. *NY Review of Books,* 37.
Ghiselin, M. T. (1974). *The Economy of Nature and the Evolution of Sex.* Berkeley: University of California Press.
Giere, R. (1985). Constructive Realism. In P. M. Churchland and C. A. Hooker (Eds.), *Images of Science.* Chicago: University of Chicago Press.
Gigerenzer, G., and K. Hug (1992). Domain-Specific Reasoning: Social Contracts, Cheating, and Perspective Change. *Cognition, 43,* 127–171.

Gilinsky, N. (1986). Species Selection as a Causal Process. *Evolutionary Biology, 20*, 248–273.
Glennan, S. (2002). Contextual Unanimity and the Units of Selection Problem. *Philosophy of Science, 69(1)*, 118–137.
Glymour, B. (1999). Population Level Causation and a Unified Theory of Natural Selection. *Biology and Philosophy, 14(4)*, 521–536.
Godfrey-Smith, P. (1992). Additivity and the Units of Selection. *PSA: Proceedings of the Biennial Meeting of the Philosophy of Science Association, 1992 Vol 1*, 315–328.
Godfrey-Smith, P., and B. Kerr (2002). Group Fitness and Multi-Level Selection: Replies to Commentaries. *Biology and Philosophy, 17(4)*, 539–550.
Godfrey-Smith, P., and R. Lewontin (1993). The Dimensions of Selection. *Philosophy of Science, 60*, 373–395.
Goldfoot, D., J. Westerborg-van Loon, W. Groeneveld, and A. Koos Slob (1980). Behavioral and Physiological Evidence of Sexual Climax in the Female Stump-Tailed Macaque (Macaca Arctoides). *Science, 208*, 1477–1479.
Goldman, J. C., J. J. McCarthy, and D. G. Peavey (1979). Growth Rate Influence the Chemical Composition of Phytoplankton in Oceanic Matters. *Nature, 279*, 210–215.
Goldman, J. C., C. D. Taylor, and P. M. Glibert (1981). Nonlinear Time-Course Uptake of Carbon and Ammonium by Marine Phytoplankton. *Marine Ecology Progress Series*, 137–148.
Gollin, E. S., G. Stahl, and E. Morgan (1989). On the Uses of the Concept of Normality in Developmental Biology and Psychology. *Advances in Child Development Behavior, 21*, 49–71.
Goodnight, C. J., and L. Stevens (1997). Experimental Studies of Group Selection: What Do They Tell Us about Group Selection in Nature? *The American Naturalist, 150*, S 59–S79.
Gould, S. J. (1982). The Meaning of Punctuated Equilibrium and Its Role in Validating a Hierarchical Approach to Macroevolution. In R. Milkman (Ed.), *Perspectives in Evolution* (pp. 83–104). Sunderland, MA: Sinauer.
Gould, S. J. (1983). The Meaning of Punctuated Equilibrium and Its Role in Validating a Hierarchical Approach to Macroevolution. In R. Milkman (Ed.), *Perspectives in Evolution*. Sunderland, MA: Sinauer.
Gould, S. J. (1987). Freudian Slip. *Natural History (Feb)*, 14–21.
Gould, S. J. (1989). Punctuated Equilibrium in Fact and Theory. *Journal of Social Biology Structure, 12*, 117–136.
Gould, S. J. (2002). *The Structure of Evolutionary Theory*. Cambridge, MA: Harvard University Press.
Gould, S. J., and N. Eldredge (1977). Punctuated Equilibria: The Tempo and Mode of Evolution Reconsidered. *Paleobiology, 3*, 115–151.
Gould, S. J., and R. C. Lewontin (1979). The Spandrels of San Marco and the Panglossian Paradigm: A Critique of the Adaptationist Programme. *Proceedings of the Royal Society of London, B205*, 581–598.
Gould, S. J., and E. S. Vrba (1982). Exaptation – A Missing Term in the Science of form. *Paleobiology, 8*, 4–15.
Grene, M. (1985). *Descartes*. Minneapolis: University of Minnesota Press.

Grene, M. (1991). *Descartes among the Scholastics*. Milwaukee: Marquette University Press.

Griesemer, J. R. (2005). The Informational Gene and the Substantial Body: On the Generalization of Evolutionary Theory by Abstraction. In N. Cartwright and M. Jones (Eds.), Idealization XII: Correcting the Model, Idealization and Abstraction in the Sciences (pp. 59–115). *Poznan Studies in the Philosophy of the Sciences and the Humanities, vol 86*. Amsterdam: Rodopi Publishers.

Griesemer, J. (2000). Development, Culture and the Units of Inheritance. *Philosophy of Science (Proceedings), 67*, S348-S368.

Griesemer, J. R., and M. J. Wade (1988). Laboratory Models, Causal Explanation, and Group Selection. *Biology and Philosophy, 3*, 67–96.

Griesemer, J. R., and W. Wimsatt (1989). Picturing Weismannism: A Case Study of Conceptual Evolution. In M. Ruse (Ed.), *What the Philosophy of Biology Is, Essays for David Hull* (pp. 75–137). Dordrecht, The Netherlands: Kluwer.

Gross, P., and N. Levitt (1994). *Higher Superstition: The Academic Left and its Quarrels with Science*. Baltimore, MD: Johns Hopkins University Press.

Hacking, I. (1983). *Representing and Intervening: Introductory Topics in the Philosophy of Natural Science*. Cambridge: Cambridge University Press.

Hacking, I. (1988). Philosophers of Experiment. *Proceedings of the Philosophy of Science Association 1988*. East Lansing, MI: Philosophy of Science Association.

Hafez, E. S. E. (1971). Reproductive Cycles. In E. S. E. Hafez (Ed.), *Comparative Reproduction of Non-human Primates*. Springfield, IL: Charles C. Thomas.

Haldane, J. B. S. (1932). *The Causes of Evolution*. London: Longmans, Green.

Hamilton, W. D. (1967). Extraordinary Sex Ratios. *Science, 156*, 477–488.

Hamilton, W. D. (1975). Innate Social Aptitudes in Man: An Approach from Evolutionary Genetics. In R. Fox (Ed.), *Biosocial Anthropology* (pp. 133–55). New York: Wiley.

Hamilton, W. D., R. Axelrod, and R. Tanese (1990). Sexual Reproduction as an Adaptation to Resist Parasites (A Review). *Proceedings of the National Academy of Sciences, U.S.A., 87*, 3566–3573.

Hampe, M., and S. R. Morgan (1988). Two Consequences of Richard Dawkins' View of Genes and Organisms. *Studies in History and Philosophy of Science, 19*, 119–138.

Hansen, T. A. (1978). Larval Dispersal and Species Longevity in Lower Tertiary Gastropods. *Science, 199*, 885–887.

Hansen, T. A. (1980). Influence of Larval Dispersal and Geographic Distribution on Species Longevity in Neogastropods. *Paleobiology, 8*, 367–377.

Haraway, D. (1989). *Primate Visions: Gender, Race, and Nature in the World of Modern Science*. New York: Routledge.

Haraway, D. (1991). Situated Knowledges: The Science Question in Feminism and the Privilege of Partial Perspective. In *Simians, Cyborgs, and Women: The Reinvention of Nature*. New York: Routledge.

Harding, S. (1986). *The Science Question in Feminism*. Ithaca, NY: Cornell University Press.

Harding, S. (Ed.). (1987). *Feminism and Methodology: Social Science Issues*. Bloomington: Indiana University Press.

Harding, S. (1989). Why "Physics" is a Bad Model for Physics. In *The End of Science? Attack and Defense (25th Nobel Conference, 1989)*. Lanham, MD: University Press of America, 1992.

Harding, S. (1991). *Whose Science? Whose Knowledge? Thinking From Women's Lives*. Ithaca, NY: Cornell University Press.

Harding, S. (1992). After the Neutrality Ideal: Science, Politics, and "Strong Objectivity." *Social Research, 59*(3), 567–587.

Harding, S. (Ed.). (1993a). *The "Racial" Economy of Science: Toward a Democratic Future*. Bloomington: Indiana University Press.

Harding, S. (1993b). Rethinking Standpoint Epistemology: What is Strong Objectivity? In L. Alcoff and E. Potter (Eds.), *Feminist Epistemologies* (pp. 49–82), New York: Routledge.

Harding, S. (1995). Strong Objectivity: A Response to the New Objectivity Question. *Synthese, 104* (3), 331–349.

Harman, G. (1965). The Inference to the Best Explanation. *Philosophical Review, 74*, 88–95.

Harman, G. (1968). Knowledge, Inference, and Explanation. *American Philosophical Quarterly, 5*, 164–173.

Haskell, T. (1990). Objectivity Is Not Neutrality: Rhetoric vs. Practice in Peter Novick's That Noble Dream. *History and Theory, 29*, 129–157.

Haslanger, S. (1993). On Being Objective and Being Objectified. In L. Antony and C. Witt (Eds.), *A Mind of One's Own: Feminist Essays on Reason and Objectivity*. Boulder, CO: Westview.

Heisler, I. L., and J. Damuth (1988). A Method for Analyzing Selection in Hierarchically Structured Populations. *American Naturalist, 130*, 582–602.

Herschel, J. F. W. (1831). *A Preliminary Discourse on the Study of Natural Philosophy.* London: Longman, Rees, Orme, Brown, and Green.

Hesse, M. (1974). *The Structure of Scientific Inference*. Berkeley: University of California Press.

Hesse, M. (1988). Socializing Epistemology. In E. McMullin (Ed.), *Construction and Constraint: The Shaping of Scientific Rationality*. South Bend, IN: University of Notre Dame Press.

Hite, S. (1976). *The Hite Report*. New York: Macmillan.

Hobbes, T. (1651, 1962). *Leviathan*, (Ed. M. Oakeshott). London: Collier Macmillan.

Hobbes. T. (1839 ff.). *Works* (Ed. Molesworth). London: Routledge.

Holden, C. (1991). Probing the Complex Genetics of Alcoholism. *Science, 251*, 163–164.

Holton, G. (1993). *Science and Anti Science*. Cambridge, MA: Harvard University Press.

Hrdy, S. B. (1986). Empathy, Polyandry, and the Myth of the Coy Female. In R. Bleier (Ed.), *Feminist Approaches to Science*. New York: Pergamon.

Hubbard. R. (1990). *The Politics of Women's Biology*. New Brunswick, NJ: Rutgers University Press.

Hubbard, R., M. S. Henifen, and B. Fried (Eds.). (1982). *Biological Woman, the Convenient Myth: A Collection of Feminist Essays and a Comprehensive Bibliography*. Cambridge, MA: Schenkman.

Hull, D. L. (1980). Individuality and Selection. *Annual Review of Ecology and Systematics, 11*, 311–332.
Hull, D. L. (1988). *Science as a Process*. Chicago: University of Chicago Press.
Hull, D. L. (1988a). Interactors versus Vehicles. In H. C. Plotkin (Ed.), *The Role of Behavior in Evolution* (pp. 19–50). Cambridge, MA: MIT Press.
Jablonski, D. (1986). Background and Mass Extinctions: The Alternation of Macroevolutionary Regimes. *Science, 231*, 129–133.
Jablonski, D. (1987). Heritability at the Species Level: Analysis of Geographic Ranges of Cretaceous Mollusks. *Science, 238*, 360–363.
Jaggar, A., and S. Bordo (Eds.). (1989). *Gender/Body/Knowledge*. New Brunswick, NJ: Rutgers University Press.
Johnson, M. (1987). *The Body in the Mind: The Bodily Basis of Meaning, Imagination, and Reason*. Chicago: University of Chicago Press.
Jukes, T. H. (1988). The Human Genome Project: Labeling Genes. *California Monthly, December*, 15–17.
Kano, T. (1980). Special Behavior of Wild Pygmy Chimpanzees (Pan Paniscus) of Wambe: A Preliminary Report. *Journal of Human Evolution, 9*, 243–260.
Kano, T. (1982). The Social Group of Pygmy Chimpanzees of Wamba. *Primates, 23*(2), 171–188.
Karjala, D. S. (1992). A Legal Research Agenda for the Human Genome Initiative. *Jurimetrics, 32*, 121–219.
Kawata, M. (1987). Units and Passages: A View for Evolutionary Biology. *Biology and Philosophy, 2*, 415–434.
Kekule, F. A. (1965). Origin of the Benzene and Structural Theory. *Chemistry, 38*.
Keller, E. F. (1985). *Reflections on Gender and Science*. New Haven, CT: Yale University Press.
Keller, E. F., and E. A. Lloyd (1992). *Keywords in Evolutionary Biology*. Cambridge, MA: Harvard University Press.
Kerr, B., and P. Godfrey-Smith (2002). Individualist and Multi-Level Perspectives on Selection in Structured Populations. *Biology and Philosophy, 17(4)*, 477–517.
Kimura, M., and T. Ohta (1971). *Theoretical Aspects of Population Genetics*. Princeton, NJ: Princeton University Press.
Kinsey, A. C., et al. (1953). *Sexual Behavior in the Human Female*. Philadelphia: W. B. Saunders.
Kirkpatrick, M. (1987). Sexual Selection by Female Choice in Polygynous Animals. *Annual Review of Ecology and Systematics, 187*, 43–70.
Kitcher, P. (1981). Explanatory Unification. *Philosophy of Science, 48*, 507–531.
Kitcher, P., K. Sterelny, and K. Waters (1990). The Illusory Riches of Sober's Monism. *Journal of Philosophy, 87*, 158–160.
Koruda, S. (1980). Social Behavior of the Pygmy Chimpanzees. *Primates, 21*(2), 181–197.
Krimbas, C. B. (1984). On Adaptation, Neo-Darwinian Tautology, and Population Fitness. *Evolutionary Biology, 17*, 1–57.
Kuhn. T. S. (1970). *The Structure of Scientific Revolutions*. Chicago: University of Chicago Press.
Kuhn, T. S. (1977). *The Essential Tension*. Chicago: University of Chicago Press.

Kuhn, T. S. (1992). *The Trouble with the Historical Philosophy of Science*. Cambridge, MA: Dept. of the History of Science, Harvard University.

Lande, R. (1980). Sexual Dimorphism, Sexual Selection, and Adaptation in Polygenic Characters. *Evolution, 34* (2), 292–305.

Lande, R., and S. J. Arnold (1983). The Measurement of Selection on Correlated characters. *Evolution, 37*, 1210–1227.

Latour, B. (1993). *We Have Never Been Modern*. Cambridge, MA: Harvard University Press.

Laudan, L. (1971). William Whewell on the Consilience of Inductions. *Monist, 55*, 368–391.

Laudan, L. (1981). A Confutation of Convergent Realism. *Philosophy of Science, 48*, 19–49. Reprinted in J. Leplin (Ed.), *Scientific Realism* (pp. 218–249). Berkeley: University of California Press.

Leacock, E. B. (1977). Women in Egalitarian Societies. In *Becoming Visible: Women in European History*. Boston: Houghton Mifflin.

Leacock, E. B. (1978a). Structuralism and Dialectics. *Reviews in Anthropology, 5*(1), 17–128.

Leacock, E. B. (1978b). Society and Gender. In *Genes and Gender*. New York: Guardian Press.

Leacock, E. B. (1981). History, Development, and the Division of Labor by Sex: Implications for Organization. *Signs, 7*(2), 474–491.

Leacock, E. B., and H. I. Safa (1986). *Women's Work: Development and the Division of Labor*. South Hadley, MA: Gender, Bergin & Garve.

Leacock, E. B., and J. Nash (1977). Ideologies of Sex: Archetypes and Stereotypes. *Annals of the New York Academy of Sciences, 285*, 618–645.

Leigh, E. G. (1977). How Does Selection Reconcile Individual Advantage with the Good of the Group? *Proceedings of the National Academy of Sciences, USA, 74*, 4542–4546.

Levay, S. (1991). A Difference in Hypothalamic Structure between Homosexual and Heterosexual Men. *Science, 253*, 1034–1037.

Levin, M. (1988). Caring New Science: Feminism and Science. *American Scholar, 57* (Winter).

Levin, B. R., and W. L. Kilmer (1974). Interdemic Selection and the Evolution of Altruism: A Computer Simulation Study. *Evolution, 28*, 527–545.

Levins, R. (1968). *Evolution in Changing Environments*. Princeton, NJ: Princeton University Press.

Levins, R., and R. C. Lewontin (1985). *The Dialectical Biologist*. Cambridge, MA: Harvard University Press.

Lévi-Strauss, C. (1956). The Family. In H. L. Shapiro (Ed.), *Man, Culture, and Society*. New York: Oxford University Press.

Lévi-Strauss, C. (1969). *The Elementary Structure of Kinship*. Boston: Beacon Press.

Lewontin, R. (1958). *Cold Spring Harbor Symposium of Quantitative Biology 22*, 395–408.

Lewontin, R. C. (1958). A General Method for Investigating the Equilibrium of Gene Frequency in a Population. *Genetics, 43*, 421–433.

Lewontin, R. C. (1962). Interdeme Selection Controlling a Polymorphism in the House Mouse. *The American Naturalist, 46* (887), 65–78.

Lewontin, R. C. (1967). The Principle of Historicity in Evolution. In P. S. Moorehead and M. M. Kaplan (Eds.), *Mathematical Challenges to the Neo-Darwinian Interpretation of Evolution* (pp. 81–88). Philadelphia: Wistar Institute Press.

Lewontin, R. C. (1970). The Units of Selection. *Annual Review of Ecology and Systematics 1*, 1–18.

Lewontin, R. C. (1974). *The Genetic Basis of Evolutionary Change*. New York: Columbia University Press.

Lewontin, R. C. (1978). Adaptation. *Scientific American, 239*, 156–169.

Lewontin, R. C. (1985). Adaptation. In R. Levins and R. C. Lewontin (Eds.), *The Dialectical Biologist* (pp. 65–84). Cambridge, MA: Harvard University Press. Originally published as "Adattamento" in *Enciclopedia Einaudi*, vol. 1. Turin, 1977.

Lewontin. R. C. (1990). *Biology as Ideology: The Doctrine of DNA*. New York: Harper Perennial.

Lewontin, R. C., and L. C. Dunn (1960). The Evolutionary Dynamics of a Polymorphism in the House Mouse. *Genetics, 45*, 701–722.

Li, C. C. (1967). Fundamental Theorem of Natural Selection. *Nature, 214*, 505–6.

Lloyd, E. (1983) A Semantic Approach to the Structure of Population Genetics. *Philosophy of Science, 51*, 242–264.

Lloyd, E. (1983). The Nature of Darwin's Support for the Theory of Natural Selection. *Philosophy of Science, 50*, 112–129.

Lloyd, E. (1986a). Thinking about Models in Evolutionary Theory. *Philosophica, 37*, 87–100.

Lloyd, E. (1986b). Empirical Evaluation of Group Selection Debates. *Philosophy of Science Association Proceedings, Vol. I, 1986*. East Lansing, MI: Philosophy of Science Association.

Lloyd, E. A. (1986c). Evaluation of Evidence in Group Selection Debates. *Proceedings of the Philosophy of Science Association 1986, 1*, 483–493.

Lloyd, E. A. (1988/1994). *The Structure and Confirmation of Evolutionary Theory*. Westport, CT: Greenwood Press. Paperback edition with new preface, Princeton University Press, 1994.

Lloyd, E. A. (1989). A Structural Approach to Defining Units of Selection. *Philosophy of Science, 56*, 395–418.

Lloyd, E. A. (1992). Unit of Selection. In E. F. Keller and E. A. Lloyd (Eds.), *Keywords in Evolutionary Biology* (pp. 334–340). Cambridge, MA: Harvard University Press.

Lloyd, E. A. (1993). Pre-Theoretical Assumptions in Evolutionary Explanations of Female Sexuality. *Philosophical Studies, 69*, 139–153.

Lloyd, E. A. (1995). Objectivity and the Double Standard for Feminist Epistemologies. *Synthese, 104* (September), 351–381.

Lloyd, E. A. (1999). Altruism Revisited. Review of *Unto Others: The Evolution and Psychology of Unselfish Behavior*, by Elliott Sober and David S. Wilson. *Quarterly Review of Biology, 74*(4), 447–449.

Lloyd, E. A. (2001). Units and Levels of Selection: An Anatomy of the Units of Selection Debates. In Rama S. Singh, Costas B. Krimbas, Diane B. Paul, John

Beatty (Eds.), *Thinking About Evolution: Historical, Philosophical, and Political Perspectives, Volume Two* (pp. 267–291). Cambridge: Cambridge University Press.

Lloyd, E. A., and S. J. Gould (1993). Species Selection on Variability. *Proceedings of the National Academy of Sciences, USA, 90*, 595–599.

Lloyd, E. A., R. C. Lewontin, and M. Feldman (2006). The Generational Cycle of State Spaces and Adequate Genetical Representation. Manuscript.

Locke. J. (1694, 1959). *Essay Concerning Human Understanding (2nd edition).* Ed. A. C. Fraser. New York: Dover.

Longino, H. (1990). *Science as Social Knowledge.* Princeton, NJ: Princeton University Press.

Longino, H. (1995). Gender, Politics, and the Theoretical Virtues. *Synthese, 104* (3), 383–397.

Longino, H. E. (1979). Evidence and Hypothesis: An Analysis of Evidential Relations. *Philosophy of Science, 46*, 35–56.

Longino, H. E. (1993a). Subjects, Power, and Knowledge: Description and Prescription in Feminist Philosophies of Science. In L. Alcoff and E. Potter (Eds.), *Feminist Epistemologies* (pp. 101–120). New York: Routledge.

Longino, H. E. (1993b). The Essential Tensions – Phase Two: Feminist. Philosophical and Social Studies of Science. In L. Antony and C. Witt (Eds.), *A Mind of One's Own: Feminist Essays on Reason and Objectivity.* Boulder, CO : Westview Press.

Longino, H. E., and R. Doell (1983). Body, Bias and Behavior: A Comparative Analysis of Reasoning in Two Areas of Biological Science. *Signs, 9*, 206–227.

MacArthur, R. H., and E. O. Wilson (1963). An Equilibrium Theory of Insular Zoogeography. *Evolution, 17*, 373–387.

MacArthur, R. H., and E. O. Wilson (1967). *The Theory of Island Biogeography.* Princeton, NJ: Princeton University Press.

Mackie, J. L. (1974). *The Cement of the Universe: A Study of Causation.* Clarendon, UK: Oxford University Press.

Mackie, J. L. (1976). *Problems from Locke.* New York: Oxford University Press.

Masters, W. H., and V. Johnson (1961). Orgasm, Anatomy of the Female. In A. Ellis and A. Abarbanal (Eds.), *Encyclopedia of Sexual Behavior, Vol. II.* New York: Hawthorn.

Masters, W. H., and V. Johnson (1966). *Human Sexual Response.* Boston: Little, Brown.

Matessi, C., and S. D. Jayakar (1976). Conditions for the Evolution of Altruism under Darwinian Selection. *Theoretical Population Biology, 9*, 360–387.

Maynard Smith, J. (1964). Group Selection and Kin Selection: A Rejoinder. *Nature, 201*, 1145–1147.

Maynard Smith, J. (1968). *Mathematical Ideas in Biology.* Cambridge: Cambridge University Press.

Maynard Smith, J. (1976). Group Selection. *Quarterly Review of Biology, 51*, 277–283.

Maynard Smith, J. (1978). *The Evolution of Sex.* Cambridge: Cambridge University Press.

Maynard Smith, J. (1981). The Evolution of Social Behavior and Classification of Models. In Kings College Sociobiology Group (Eds.), *Current Problems in Sociobiology* (pp. 29–44). Cambridge: Cambridge University Press.

Maynard Smith, J. (1984). The Population as a Unit of Selection. In B. Shorrocks (Ed.), *Evolutionary Ecology, 23rd British Ecological Society Symposium* (pp. 195–202). Oxford: Blackwell.

Maynard Smith, J. (1987). Evolutionary Progress and Levels of Selection. In J. Dupre (Ed.), *The Latest on the Best*. Cambridge, MA: MIT Press.

Mayo, D., and N. Gilinsky (1987) *Philosophy of Science, 54*, 515–538.

Mayr, E. (1963). *Animal Species and Evolution*. Cambridge, MA: Harvard University Press.

Mayr, E. (1967). Evolutionary Challenges to the Mathematical Interpretation of Evolution. In P. S. Moorehead and M. M. Kaplan (Eds.), *Mathematical Challenges to the Neo-Darwinian Interpretation of Evolution* (pp. 47–54). Philadelphia: Wistar Institute Press.

Mayr, E. (1976). *Evolution and the Diversity of Life*. Cambridge, MA: Harvard University Press.

Mayr, E. (1978). Evolution. *Scientific American, 239*, 49–55.

Mayr, E. (1982a). Adaptation and Selection. *Biologisches Zentralblatt, 101*, 161–174.

Mayr, E. (1982b). *The Growth of Biological Thought*. Cambridge, MA: Harvard University Press.

Mayr, E. (1983). How to Carry Out the Adaptationist Program? *American Naturalist, 121*, 324–334.

McDowell, J. (1979). Virtue and Reason. *Monist, 62*, 331–350.

McDowell, J. (1988). Values and Secondary Qualities. In G. Sayre-McCord (Ed.), *Essays on Moral Realism*. Ithaca, NY: Cornell University Press.

McKusick, V. A. (1989). Mapping and Sequencing the Human Genome. *New England Journal of Medicine, 320*, 910–915.

McMullin, E. (Ed.). (1988a). *Construction and Constraint: The Shaping of Scientific Rationality*. South Bend, IN: University of Notre Dame Press.

McMullin, E. (1988b). Panel Discussion. In E. McMullin (Ed.), *Construction and Constraint: The Shaping of Scientific Rationality*. South Bend, IN: University of Notre Dame Press.

Michod, R. (1999). *Darwinian Dynamics*. Princeton, NJ: Princeton University Press.

Michod, R., and B. Levin (Eds.). (1988). *The Evolution of Sex: An Examination of Current Ideas*. Sunderland, MA: Sinauer.

Mill, J. S. (1859, 1989). *On Liberty*. Cambridge: Cambridge University Press.

Miller, G. F., and P. M. Todd (1994). A Bottom-Up Approach with a Clear View of the Top: How Evolutionary Psychology Can Inform Adaptive Behavior Research. *Adaptive Behavior, 3*(1), 83–95.

Mitchell, S. (1987). Competing Units of Selection? A Case of Symbiosis. *Philosophy of Science, 54*, 351–367.

Mithin, S. (Spring 1997). Review of *The Adapted Mind. Journal of Anthropological Research, 53*, 100–102.

Mori, A. (1984). An Ethological Study of Pygmy Chimpanzees in Wambe Zaire: A Comparison with Chimpanzees. *Primates, 25*(3), 255–278.
Morris, D. (1967). *The Naked Ape*. London: Jonathan Cape.
Moulines, C. U. (1975). A Logical Reconstruction of Simple Equilibrium Thermodynamics. *Erkenntnis, 9*, 101–130.
Munson, R. (1971). Biological Adaptation. *Philosophy of Science, 38*, 200–215.
Murphy, J. P. (1990). *Pragmatism: From Peirce to Davidson*. Boulder, CO: Westview Press.
Nagel, T. (1979). *Mortal Questions*. Cambridge: Cambridge University Press.
Nagel, T. (1980). The Limits of Objectivity. In *The Tanner Lectures on Human Values* (pp. 77–139). Salt Lake City: University of Utah Press.
Nagel, T. (1986). *The View from Nowhere*. New York: Oxford University Press.
Nagel, T. (1993, March 4). The Mind Wins! Review of J. R. Searle's *The Rediscovery of the Mind. New York Review of Books*, 37–41.
Nelson, L. H. (1990). *Who Knows: From Quine to a Feminist Empiricism*. Philadelphia: Temple University Press.
Nelson, L. H. (1993). Epistemological Communities. In L. Alcoff and E. Potter (Eds.), *Feminist Epistemologies* (pp. 121–160). New York: Routledge.
Newton, I. (1726, 1972). *Philosophiae Naturalis Principia Mathematica* (3rd edition). A. Koyre and I. B. Cohen (Eds.). Cambridge, MA: Harvard University Press.
Nicholson, A. J. (1960). The Role of Population Dynamics in Natural Selection. In Tax, D. (Ed.), *Evolution After Darwin: The Evolution of Life* (pp. 477–522). Chicago: University of Chicago Press.
Nietzsche, F. (1887, 1967). *On the Genealogy of Morals & Ecco Homo*. New York: Vintage Books, Random House.
Nisbett, R. E., and P. W. Cheng (1988). Conditional Reasoning. 29th Annual Meeting of the Psychonomic Society, Chicago.
Nisbett, R. E., G. T. Fong, D. Lehman, and P. W. Cheng (1987). Teaching Reasoning. *Science, 238*, 625–631.
Novick, P. (1988). *That Noble Dream: The "Objectivity Question" and the American Historical Profession*. Cambridge: Cambridge University Press.
Nowotny, H., and H. Rose (Eds.). (1979). *Counter-Movements in the Sciences*. Dordrecht, The Netherlands: Reidel.
Nunney, L. (1985). Group Selection, Altruism, and Structured-Deme Models. *American Naturalist, 126*, 212–230.
Ohm, K. (1983). Hierarchical Theory of Selection: The Covariance Formula of Selection and Its Application. *Bulletin of the Biometrical Society of Japan, 4*, 25–33.
Okasha, S. (2004). Multilevel Selection and the Partitioning of Covariance: A Comparison of Three Approaches. *Evolution, 58*(3), 486–494.
Oster, G. F., and E. O. Wilson (1978). *Caste and Ecology in the Social Insects*. Princeton, NJ: Princeton University Press.
Peirce, C. S. (1868, 1992). Some Consequences of Four Incapacities. *Journal of Speculative Philosophy, 2*, 140–157. Reprinted in N. Houser and C. Kloesel (Eds.), *The Essential Peirce Vol. 1*. Bloomington: Indiana University Press.

Peirce, C. S. (1871, 1992). Review of Fraser's *The Works of George Berkeley. North American Review, 113*, 449–472. Reprinted in N. Houser and C. Kloesel (Eds.), *The Essential Peirce Vol. 1*. Bloomington: Indiana University Press.

Peirce. C. S. (1877, 1992). The Fixation of Belief. *Popular Science Monthly, 12*, 1–15. Reprinted in N. Houser and C. Kloesel (Eds.), *The Essential Peirce Vol. 1*. Bloomington: Indiana University Press.

Peirce, C. S. (1878a, 1992). How to Make our Ideas Clear. *Popular Science Monthly, 12*, 286–302. Reprinted in N. Houser and C. Kloesel (Eds.), *The Essential Peirce Vol. 1*. Bloomington: Indiana University Press.

Peirce, C. S. (1878b, 1992). The Order of Nature. *Popular Science Monthly, 13*, 203–217. Reprinted in N. Houser and C. Kloesel (Eds.), *The Essential Peirce Vol. 1*. Bloomington: Indiana University Press.

Peirce, C. S. (1892, 1992). The Doctrine of Necessity Examined. *The Monist, 2*, 321–327. Reprinted in N. Houser and C. Kloesel (Eds.), *The Essential Peirce Vol. 1*. Bloomington: Indiana University Press.

Perutz, M. F. (1995, December 21).The Pioneer Defended. Review of Gerald L. Geison's *The Private Science of Louis Pasteur* (Princeton, NJ: Princeton University Press). *New York Review of Books, XLII* (20).

Pfeifer, E. J. (1972). United States. In Glick (Ed.), *The Comparative Reception of Darwinism* (pp. 185–210). Austin: University of Texas Press.

Pickering, A. (1984). *Constructing Quarks: A Sociological History of Particle Physics*. Edinburgh: Edinburgh University Press.

Pinker, S. (1997). *How the Mind Works*. New York: W. W. Norton.

Pinker, S., and P. Bloom (1990). Natural Language and Natural Selection. *Behavioral and Brain Sciences, 13*, 707–784. Reprinted (1992) in J. Barkow, L. Cosmides, and J. Tooby (Eds.), *The Adapted Mind*, pp. 451–493.

Platt, J. R. (1964). Strong Inference. *Science, 146*, 347–353.

Polanyi, M. (1969a). The Republic of Science: Its Political and Economic Theory. In *Knowing and Being: Essays by Michael Polanyi*. Chicago: University of Chicago Press.

Polanyi, M. (1969b). The Structure of Consciousness. In *Knowing and Being: Essays by Michael Polanyi*. Chicago: University of Chicago Press.

Potter, E. (1993). Gender and Epistemic Negotiation. In L. Alcoff and E. Potter (Eds.), *Feminist Epistemologies* (pp. 161–186). Routledge, New York.

Price, G. R. (1972). Extension of Covariance Selection Mathematics. *Annals of Human Genetics, 35*, 485–490.

Proctor, R. (1991). *Value-Free Science? Purity and Power in Modern Knowledge*. Cambridge, MA: Harvard University Press.

Pugh, G. (1977). *Biological Origins of Human Values*. New York: Basic Books.

Putnam, H. (1981). *Reason, Truth and History*. Cambridge: Cambridge University Press.

Putnam, H. (1988). *Representation and Reality*. Cambridge, MA: MIT Press.

Putnam, H. (1992). *Renewing Philosophy*. Cambridge, MA: Harvard University Press.

Quine, W. V. (1960). *Word and Object*. Cambridge, MA: Harvard University Press.

Quine, W. V. (1963). Two Dogmas of Empiricism. In *From a Logical Point of View* (Revised Edition) (pp. 20–46). New York: Harper & Row.

Quine, W. V. (1969). Epistemology Naturalized. In *Ontological Relativity and Other Essays*. New York: Columbia University Press.

Quine, W. V. (1974). *The Roots of Reference*. La Salle, IL: Open Court.

Quine, W. V. (1981). *Theories and Things*. Cambridge, MA: Harvard University Press.

Quinn, J. F., and A. E. Dunham (1983). On Hypothesis Testing in Ecology and Evolution. *American Naturalist, 122*, 602–617.

Ranelagh, J. (1986). *The Agency: The Rise and Decline of the CIA*. New York: Simon and Schuster.

Reid, J. P. (1970). *A Law of Blood: The Primitive Law of the Cherokee Nation*. New York: New York University Press.

Reiter, R. R. (Ed.). (1975). *Toward an Anthropology of Women*. New York: Monthly Review Press.

Rescher, N. (1978). *Peirce's Philosophy of Science: Critical Studies in His Theory of Induction and Scientific Method.* South Bend, IN: University of Notre Dame Press.

Rescher, N. (1992). A System of Pragmatic Idealism, Vol. I: Human Knowledge. In *Idealist Perspective*. Pittsburgh: University of Pittsburgh Press.

Roan, S. (1991, April 18). Check It Out. *Oakland Tribune*.

Rodis-Lewis, G. (1993). *From Metaphysics to Physics: Essays on the Philosophy and Science of Rene Descartes*. New York: Oxford University Press.

Rorty, R. (1979). *Philosophy and the Mirror of Nature*. Princeton, NJ: Princeton University Press.

Rorty, R. (1982). *Consequences of Pragmatism: Essays, 1972–1980*. Minneapolis: University of Minnesota Press.

Rorty, R. (1986). Pragmatism, Davidson and Truth. In E. LePore (Ed.), *Truth and Interpretation: Perspectives on the Philosophy of Donald Davidson*. Oxford: Basil Blackwell.

Rorty, R. (1987). Pragmatism and Philosophy. In K. Baynes, J. Bohman, and T. McCarthy (Eds.), *After Philosophy: End or Transformation?* Cambridge, MA: MIT Press.

Rorty, R. (1988). Is Natural Science a Natural Kind? In E. McMullin (Ed.), *Construction and Constraint: The Shaping of Scientific Rationality*. South Bend, IN: University of Notre Dame Press.

Rorty, R. (1989). *Contingency, Irony and Solidarity*. Cambridge: Cambridge University Press.

Rose, H. (1983). Hand, Brain and Heart: A Feminist Epistemology for the Natural Sciences. *Signs, 9*, 73–90.

Roughgarden, J. (1983). Competition and Theory in Community Ecology. *American Naturalist, 122*, 583–601.

Rouse, J. (1987). *Knowledge and Power: Toward a Political Philosophy of Science*. Ithaca, NY: Cornell University Press.

Rudwick, M. J. S. (1976). *The Meaning of Fossils* (2nd ed.). New York: Neale Watson Academic Publishers.

Ruse, M. (1975a). Charles Darwin's Theory of Evolution: An Analysis. *Journal of the History of Biology, 8*(2), 219–241.

Ruse, M. (1975b). Darwin's Debt to Philosophy. *Studies in the History and Philosophy of Science, 6*(2), 159–181.

Ruse, M. (1979). *The Darwinian Revolution: Science Red in Tooth and Claw*. Chicago: University of Chicago Press.

Ruse, M. (1981). *Is Science Sexist? And Other Problems in the Biomedical Sciences*. Dordrecht: Reidel.

Sarkar, S. (1994). The Selection of Alleles and the Additivity of Variance. *PSA: Proceedings of the Biennial Meeting of the Philosophy of Science Association, 1994, Vol. One*, 3–12.

Scheffler, I. (1967). *Science and Subjectivity*. Indianapolis: Bobbs-Merrill.

Searle, J. R. (1984). *Minds, Brains and Science.* Cambridge, MA: Harvard University Press.

Searle, J. R. (1992). *The Rediscovery of the Mind*. Cambridge, MA: MIT Press.

Searle, J. R. (1993). Rationality and Realism: What Is at Stake? *Daedalus, 122*(4), 55–84.

Sedgwick, P. (1973). Illness – Mental and Otherwise. *Hastings Center Report, 1*, 19–40.

Sellars. W. (1968). *Science and Metaphysics: Variations on Kantian Themes*. London: International Library of Philosophy and Scientific Method, Routledge & Kegan Paul.

Shanahan, T. (1997). Pluralism, Antirealism, and the Units of Selection. *Acta Biotheoretica, 45*, 117–126.

Shapere, D. (1981). Meaning and Scientific Change. In Ian Hacking (Ed.), *Scientific Revolutions* (pp. 28–59). New York: Oxford University Press.

Shapin, S., and S. Schaffer. (1985). *Leviathan and the Air Pump: Hobbes, Boyle, and the Experimental Life*. Princeton, NJ: Princeton University Press.

Shepard, R. N. (1987). Evolution of a Mesh between Principles of the Mind and Regularities of the World. In J. Dupre (Ed.), *The Latest on the Best: Essays on Evolution and Optimality*. Oxford: Blackwell.

Shimony, A. (1970). Scientific Inference. In R. Colodny (Ed.), *The Nature and Function of Scientific Theories*. Pittsburgh: University of Pittsburgh Press.

Simberloff, D. (1983). Competition Theory, Hypothesis Testing, and Other Community Ecological Buzzwords. *American Naturalist, 122*, 625–635.

Simberloff D., and E. O. Wilson (1969). Experimental Zoogeography of Islands: The Colonization of Empty Islands. *Ecology, 50*, 278–296.

Simberloff, D., and E. O. Wilson (1970). Experimental Zoogeography of Islands: A Two-Year Record of Colonization. *Ecology, 51*, 934–934.

Simpson, G. G. (1953). *The Major Features of Evolution*. New York: Columbia University Press.

Singer, I., and J. Singer (1972). Periodicity of Sexual Desire in Relation to Time of Ovulation in Women. *Journal of Biosocial Science, 4*, 471–481.

Siskind, J. (1978). Kinship and Mode of Production. *American Anthropologist, 80*.

Slatkin, M. (1972). On Treating the Chromosome as the Unit of Selection. *Genetics, 72*, 157–168.

Slatkin, M. (1981). A Diffusion Model of Species Selection. *Paleobiology, 7* (4), 421–425.

Slatkin, M., and M. Wade (1978). Group Selection on a Quantitative Character. *Proceedings of the National Academy of Sciences, USA, 75*, 3531–3534.

Slobodkin, L. B., and A. Rapoport (1974). An Optimal Strategy of Evolution. *Quarterly Review of Biology, 49*, 181–200.

Smart, J. J. C. (1963). *Philosophy and Scientific Realism*. New York: Routledge & Kegan Paul.

Smith, L., and L. Hood (1987). Mapping and Sequencing the Human Genome: How to Proceed. *Biotechnology, 5*, 933–939.

Sneed, J. (1971). *The Logical Structure of Mathematical Physics*. Dordrecht: Reidel.

Sober, E. (1981). Holism, Individualism, and the Units of Selection. *Proceedings of the Philosophy of Science Association 1980, 2*, 93–121.

Sober, E. (1984). *The Nature of Selection*. Cambridge, MA: MIT Press.

Sober, E. (1990). The Poverty of Pluralism: A Reply to Sterelny and Kitcher. *The Journal of Philosophy, 87*(3),151–158.

Sober, E., and R. C. Lewontin (1982). Artifact, Cause and Genic Selection. *Philosophy of Science, 49*, 157–180.

Sober, E., and D. S. Wilson (1994). A Critical Review of Philosophical Work on the Units of Selection Problem. *Philosophy of Science, 61*, 534–555.

Sober, E., and D. S. Wilson (1998). *Unto Others: The Evolution and Psychology of Unselfish Behavior*. Cambridge, MA: Harvard University Press.

Spencer, H. G. (2001). Optimization, Limitations of. In N. J. Smelser and P. B. Baltes (Eds.), *International Encyclopedia of the Social and Behavioral Sciences*. Oxford: Elsevier.

Spencer, H., and J. Masters (1992). Sexual Selection: Contemporary Debates. In E. F. Keller and E. A. Lloyd (Eds.), *Keywords in Evolutionary Biology*. Cambridge, MA: Harvard University Press.

Stanford, P. K. (2001). The Units of Selection and the Causal Structure of the World. *Erkenntnis, 54*, 215–233.

Stanley, S., and X. Yang (1987). Approximate Evolutionary Stasis for Bivalve Morphology Over Millions of Years: A Multivariate, Multilineage Study. *Paleobiology, 13*, 113–139.

Stanley, S. (1975a). Clades versus Clones in Evolution: Why We Have Sex. *Science, 190*, 382–383.

Stanley, S. (1975b). A Theory of Evolution above the Species Level. *Proceedings of the National Academy of Sciences USA, 72* (2), 646–650.

Stanley, S. (1979). *Macroevolution: Pattern and Process*. New York: Freeman.

Starfield, A. M., et al. (1980). An Exploratory Model of Impala Population Dynamics. In S. Levin (Ed.), *Mathematical Modelling in Biology and Ecology*, Vol. 33 of Lecture Notes in *Biomathematics*. Berlin and New York: Springer-Verlag.

Stegmuller, W. (1976). *The Structure and Dynamics of Theories*. New York: Springer-Verlag.

Stein, H. (1993). On Philosophy and Natural Philosophy in the Seventeenth Century. In P. French, T. Uehling, Jr., and H. Wettstein (Eds.), *Midwest Studies in Philosophy, Volume XVIII*: Philosophy of Science, pp. 177–201.

Sterelny, K., and P. Kitcher (1988). The Return of the Gene. *Journal of Philosophy, 85*, 339–361.

Stich, S. (1983). *From Folk Psychology to Cognitive Science: The Case against Belief*. Cambridge, MA: MIT Press.

Strawson, P. (1959). *Individuals*. London: Methuen.

Strong, D. R. (1983). Natural Variability and the Manifold Mechanisms of Ecological Communities. *American Naturalist, 122*, 636–660.

Stroud, B. (1980). Berkeley v. Locke on Primary Qualities. *Philosophy, 55*, 149–166.

Stroud. B. (1984). *The Significance of Philosophical Scepticism*. New York: Oxford University Press.

Strum, S. (1987). *Almost Human: A Journey into the World of Baboons*. New York: Random House.

Suppe, F. (1972). What's Wrong with the Received View on the Structure of Scientific Theories? *Philosophy of Science, 39*, 1–19.

Suppe, F. (1973). Theories, Their Formulations, and the Operational Imperative. *Synthese, 25*, 129–164.

Suppe, F. (1974). Theories and Phenomena. In W. Leinfellner and E. Kohler (Eds.), *Developments of the Methodology of Social Science* (pp. 45–92). Dordrecht: Reidel.

Suppe, F. (1974a). Some Philosophical Problems in Biological Speciation and Taxonomy. In Wojcieckowski (Ed.), *Conceptual Basis of the Classification of Knowledge* (pp. 190–243). Munich: Verlag Dokumentation.

Suppe, F. (1976). Theoretical Laws. In M. Przelecki et al. (Eds.), *Formal Methods in the Methodology of Empirical Sciences* (pp. 247–267). Dordrecht, Boston: Reidel.

Suppe, F. (Ed.) (1977). *The Structure of Scientific Theories* (2nd ed.). Urbana: University of Illinois Press.

Suppe, F. (1979). Theory Structure. In *Current Research in Philosophy of Science* (pp. 317–338). East Lansing, MI: Philosophy of Science Association.

Suppes, P. (1957). *Introduction to Logic*. Princeton, NJ: D. Van Nostrand and Co.

Suppes, P. (1961). A Comparison of the Meaning and Use of Models in Mathematics and the Empirical Sciences. In H. Freudenthal (Ed.), *The Concept and the Role of the Model in Mathematics and Natural and Social Sciences* (pp. 163–177). Dordrecht: Reidel.

Suppes, P. (1962). Models of Data. In E. Nagel, P. Suppes, and A. Tarski (Eds.), *Logic, Methodology, and the Philosophy of Science*. Proceedings of the 1960 International Congress, Vol. 1 (pp. 252–261). Stanford, CA: Stanford University Press.

Suppes, P. (1967). What Is a Scientific Theory? In S. Morgenbesser (Ed.), *Philosophy of Science Today*. New York: Meridian Books.

Suppes, P. (1978). The Plurality of Science. *Proceedings of the Philosophy of Science Association*. East Lansing, MI: Philosophy of Science Association.

Symons, D. (1987). If We're All Darwinians, What's the Fuss About? In C. Crawford, D. Krebs, and M. Smith (Eds.), *Sociobiology and Psychology.* Hillsdale, NJ: Lawrence Erlbaum Associates.

Teller, P. (1992). Subjectivity and Knowing What It's Like: Emergence or Reduction? In J. Kim (Ed.), *Essays on the Prospects of Nonreductive Physicalism.* Berlin, GR: Walter de Gruyter.

Temkin, O. (1977). *The Double Face of Janus.* Baltimore, MD: Johns Hopkins University Press.

Thagard, P. (1978). The Best Explanation: Criteria for Theory Choice. *Journal of Philosophy, 75,* 78–92.

Thagard, P. (1992). *Conceptual Revolutions.* Princeton, NJ: Princeton University Press.

Thoday, J. M. (1953). Components of Fitness. *Symposium of the Society for Experimental Biology, 7,* 96–113.

Thompson, P. (1983). The Structure of Evolutionary Theory: a Semantic Approach. *Studies in History and Philosophy of Science, 14,* 215–229.

Thompson, P. (1985). Sociobiological Explanation and the Testability of Sociobiological Theory. In James H. Fetzer (Ed.), *Sociobiology and Epistemology.* Dordrecht: Reidel.

Toft, C. A., and P. J. Shea (1983). Detecting Communitywide Patterns: Estimating Power Strengthens Statistical Inference. *American Naturalist, 122,* 618–625.

Tooby, J. (1985). The Emergence of Evolutionary Psychology. In D. Pines (Ed.), *Emerging Syntheses in Science. Proceedings of the Founding Workshops of the Santa Fe Institute* (pp. 67–75). Santa Fe, NM: The Santa Fe Institute.

Tooby, J., and L. Cosmides (1989). Evolutionary Psychology and the Generation of Culture, Part I: Theoretical Considerations. *Ethology and Sociobiology, 10,* 29–49. Pt II: Case Study: A Computational Theory of Social Exchange. *Ethology and Sociobiology, 10,* 51–97.

Tooby, J., and L. Cosmides (1990a). The Past Explains the Present: Emotional Adaptations and the Structure of Ancestral Environments. *Ethology and Sociobiology, 11,* 375–424.

Tooby, J., and L. Cosmides (1990b). On the Universality of Human Nature and the Uniqueness of the Individual: The Role of Genetics and Adaptation. *Journal of Personality*, 58, 17–67.

Tooby, J., and L. Cosmides (1992). The Psychological Foundations of Culture. In J. Barkow, L. Cosmides, and J. Tooby (Eds.), T*he Adapted Mind: Evolutionary Psychology and the Generation of Culture* (pp. 19–136). New York: Oxford University Press.

Tooby, J., and I. DeVore (1987). The Reconstruction of Hominid Behavioral Evolution through Strategic Modeling. In W. G. Kinzey (Ed.), *The Evolution of Human Behavior: Primate Models* (pp. 183–237). New York: State University of New York Press.

Traweek, S. (1988). *Beamtimes and Lifetimes: The World of High Energy Physicists.* Cambridge, MA: Harvard University Press.

Trivers, R. (1971). The Evolution of Reciprocal Altruism. *Quarterly Review of Biology, 46,* 35–57.

Tuana, N. (Ed.). (1989). *Feminism and Science*. Bloomington: Indiana University Press.
Tuana, N. (1995). The Values of Science: Empiricism from a Feminist Perspective. *Synthese, 104*(3), 441–461.
Turner, J. (1986). In D. M. Raup and D. Jablonski (Eds.), *Patterns and Processes in the History of Life* (pp. 183–207). Berlin: Springer.
Uyenoyama, M. K. (1979). Evolution of Altruism under Group Selection in Large and Small Populations in Fluctuating Environments. *Theoretical Population Biology, 15*, 58–85.
Uyenoyama, M. K., and M. W. Feldman (1980). Evolution of Altruism under Group Selection in Large and Small Populations in Fluctuating Environments. *Theoretical Population Biology, 17*, 380–414.
Van der Steen, W. J. and H. A. Van den Berg (1999). Dissolving Disputes over Genic Selectionism. *Journal of Evolutionary Biology, 12*, 184–187.
van Fraassen, B. C. (1969). Meaning Relations and Modalities. *Nous, 3*, 155–167.
van Fraassen, B. C. (1970). On the Extension of Beth's Semantics of Physical Theories. *Philosophy of Science, 37*, 325–339.
van Fraassen, B. C. (1972). A Formal Approach to the Philosophy of Science. In R. E. Colodny (Ed.), *Paradigms and Paradoxes*. Pittsburgh: University of Pittsburgh Press.
van Fraassen, B. C. (1974). The Labyrinth of Quantum Logic. In R. S. Cohen and M. Wartofsky (Eds.), *Logical and Epistemological Studies in Contemporary Physics*. Boston Studies in the Philosophy of Science, Vol XIII. Dordrecht, Netherlands: D. Reidel Publishing Company.
van Fraassen, B. C. (1980). *The Scientific Image*. Oxford: Clarendon Press.
van Fraassen, B. C. (1989). *Laws and Symmetry*. Oxford: Oxford University Press.
van Fraassen, B. C. (1986). Aim and Structure of Scientific Theories. In R. B. Marcus, G. Dorn, and P. Weingartner (Eds.), *Proceedings of the Seventh International Congress of Logic, Methodology, and Philosophy of Science* (pp. 397–318). Amsterdam: North-Holland Publishing Company.
Vehrencamp, S. L., and J. W. Bradbury (1984). Mating Systems and Ecology. In J. R. Krebs (Ed.), *Behavioural Ecology: An Evolutionary Approach* (pp. 251–278). Sunderland, MA: Sinauer.
Vicedo, M. (MS, 1991). *The History, Scientific Value, and Social Implications of the Human Genome Project*.
Voss, S. (Ed.) (1993). *Essays on the Philosophy and Science of Rene Descartes*. New York: Oxford University Press.
Vrba, E. (1980). Evolution, Species and Fossils: How Does Life Evolve? *South African Journal of Science, 76*, 61–84.
Vrba, E. (1983). Macroevolutionary Trends: New Perspectives on the Roles of Adaptation and Incidental Effect. *Science, 221*, 387–389.
Vrba, E. (1984). What Is Species Selection? *Systematic Zoology, 33*, 318–328.
Vrba, E. (1989). Levels of Selection and Sorting with Special Reference to the Species Level. *Oxford Surveys in Evolutionary Biology, 6*, 111–168.
Vrba, E., and N. Eldredge (1984). Individuals, Hierarchies and Processes: Towards a More Complete Evolutionary Theory. *Paleobiology, 10*, 146–171.

Vrba, E., and S. J. Gould (1986). The Hierarchical Expansion of Sorting and Selection: Sorting and Selection Cannot Be Equated. *Paleobiology, 12*, 217–228.

Waddington, C. H. (1956). Genetic Assimilation of the Bithorax Phenotype. *Evolution, 10*, 1–13.

Wade, M. J. (1976). Group Selection Among Laboratory Populations of Tribolium. *Proceedings of the National Academy of Sciences USA, 73*(12), 4604–4607.

Wade, M. J. (1977). An Experimental Study of Group Selection. *Evolution, 31*, 134–153.

Wade, M. J. (1978). A Critical Review of the Models of Group Selection. *Quarterly Review of Biology, 53*, 101–114.

Wade, M. J. (1980). Kin Selection: Its Components. *Science, 210*, 665–667.

Wade, M. J. (1985). Soft Selection, Hard Selection, Kin Selection, and Group Selection. *American Naturalist, 125*, 61–73.

Wade, M. J., and F. Breden (1981). The Effect of Inbreeding on the Evolution of Altruistic Behavior by Kin Selection. *Evolution, 35*, 844–858.

Wade, M. J., and McCauley, D. E. (1980). Group Selection: The Phenotypic and Genotypic Differentiation of Small Populations. *Evolution, 34*, 799–812.

Waters, K. (1986). *Models of Natural Selection: From Darwin to Dawkins*. Ph.D. diss., History and Philosophy of Science Department. Bloomington: Indiana University.

Waters, K. (1991). Tempered Realism About the Force of Selection. *Philosophy of Science 58*(4), 553–573.

Watson, J. D. (1990). The Human Genome Project: Past, Present, and Future. *Science, 248*, 44.

Wessels, L. (1976). Laws and Meaning Postulates (in van Fraassen's View of Theories). In R. S. Cohen et al. (Eds.), *PSA 1974* (pp. 215–235). Dordrecht: Reidel.

Whitbeck, C. (1984). A Different Reality: Feminist Ontology. In C. Gould (Ed.), *Beyond Domination: New Perspectives on Women and Philosophy*. Totowa, NJ: Rowman & Allenheld.

Wiggins, D. (1976). Truth, Invention and the Meaning of Life. *Proceedings of the British Academy, LXII*, 331–378.

Wiggins, D. (1980). *What Would Be a Substantial Theory of Math? Philosophical Subjects: Essays Presented to P. F. Strawson*. New York: Oxford University Press.

Wiggins, D. (1987). *Needs, Values, Truth: Essays in the Philosophy of Value*. Oxford: Basil Blackwell.

Williams, B. (1978). *Descartes: The Project of Pure Enquiry*. Atlantic Highlands, NJ: Humanities Press.

Williams, B. (1981). *Moral Luck: Philosophical Papers, 1973–1980*. Cambridge: Cambridge University Press.

Williams, B. (1985). *Ethics and the Limits of Philosophy*. Cambridge, MA: Harvard University Press.

Williams, G. C. (1966). *Adaptation and Natural Selection*. Princeton, NJ: Princeton University Press.

Williams, G. C. (1975). *Sex and Evolution*. Princeton, NJ: Princeton University Press.

Williams, G. C. (1985). A Defense of Reductionism in Evolutionary Biology. *Oxford Surveys in Evolutionary Biology, 2*, 1–27.
Williams, G. C. (1992). *Natural Selection: Domains, Levels, and Challenges*. New York: Oxford University Press.
Williams, M. B. (1982). The Importance of Prediction Testing in Evolutionary Biology. *Erkenntnis, 17*, 291–306.
Wilson, D. S. (1975). A General Theory of Group Selection. *Proceedings of the National Academy of Sciences, USA, 72*,143–146.
Wilson, D. S. (1980). *The Natural Selection of Populations and Communities*. Menlo Park, CA: Benjamin Cummings.
Wilson, D. S. (1983). Group Selection Controversy: History and Current Status. *Annual Review of Ecology and Systematics, 14*, 159–187.
Wilson, D. S., and R. K. Colwell (1981). Evolution of Sex Ratio in Structured Demes. *Evolution, 35*, 882–897.
Wilson, E. O. (1975). *Sociobiology*. Cambridge, MA: Harvard University Press.
Wilson, E. O., and Simberloff, D. S. (1969). Experimental Zoogeography of Islands: Definition and Monitoring Techniques. *Ecology, 50*, 267–278.
Wilson, M. D. (1978). *Descartes*. London: Routledge & Kegan Paul.
Wilson, M. D. (1979). Superadded Properties: The Limits of Mechanism in Locke. *American Philosophical Quarterly, 16*, 143–150.
Wilson, M. D. (1982). Did Berkeley Completely Misunderstand the Basis of the Primary-Secondary Qualities? In C. Thrbayne (Ed.), *Berkeley: Critical and Interpretive Essays* (pp. 162–176). Minneapolis: University of Minnesota Press.
Wilson, M. D. (1993). Descartes on the Perception of Primary Qualities. In S. Voss (Ed.), *Essays on the Philosophy and Science of Rene Descartes*. New York: Oxford University Press.
Wilson, R. A. (2003). Pluralism, Entwinement, and the Levels of Selection. *Philosophy of Science, 70(3)*, 531–552.
Wimmer, H., and J. Pemer (1983). Beliefs about Beliefs: Representation and Constraining Function of Wrong Beliefs in Young Children's Understanding of Deception. *Cognition, 13*, 103–128.
Wimsatt, W. (1980). Reductionist Research Strategies and Their Biases in the Units of Selection Controversy. In T. Nickles (Ed.), *Scientific Discovery: Case Studies* (pp. 213–259). Dordrecht, The Netherlands: Reidel.
Wimsatt, W. (1981). Units of Selection and the Structure of the Multi-Level Genome. *Proceedings of the Philosophy of Science Association 1980, 2*, 122–183.
Wittgenstein, L. (1953). *Philosophical Investigations*. Oxford: Blackwell.
Wittgenstein, L. (1965). A Lecture on Ethics. *Philosophical Review, 74*, 3–11.
Wolfe, L. (1979). Behavioral Patterns of Estrous Females of the Arachiyama West Troop of Japanese Macaques (Macaca Fuscata). *Primates, 20*(4), 525–534.
Wolpert, L. (1992). *The Unnatural Nature of Science: Why Science Does Not Make (Common) Sense*. Cambridge, MA: Harvard University Press.
Wrangham, R. W., W. C. McGrew, and F. B. M. de Waal (Eds.). (1994). *Chimpanzee Cultures*. Cambridge, MA: Harvard University Press.
Wright, C. (1992). *Truth and Objectivity*. Cambridge, MA: Harvard University Press.
Wright, S. (1931). Evolution in Mendelian Populations. *Genetics, 10*, 97–159.

Wright, S. (1945). Tempo and Mode in Evolution: A Critical Review. *Ecology, 26*, 415–419.

Wright, S. (1980). Genic and Organismic Selection. *Evolution, 34*, 825–843.

Wylie, A. (1988). Methodological Essentialism: Comments on Philosophy, Sex and Feminism. *Atlantis, 13*(2), 11–14.

Wylie, A. (1989). Archaeological Cables and Tacking: The Implications of Practice for Bernstein's Beyond Objectivism and Relativism. *Philosophy of Social Science, 19*, 1–18.

Wynne-Edwards, V. C. (1962). *Animal Dispersion in Relation to Social Behavior.* Edinburgh: Oliver and Boyd.

Ziman, J. (1968). *Public Knowledge: An Essay Concerning the Social Dimension of Science.* London: Cambridge University Press.

Zinder, N. D. (1990). The Genome Initiative: How to Spell Human. *Scientific American, 96*, July.

Index

Printed in the United States
By Bookmasters